Newnes
PC Troubleshooting
Pocket Book

Newnes PC Troubleshooting Pocket Book

Second edition

Howard Anderson
and Mike Tooley

AMSTERDAM BOSTON HEIDELBERG LONDON NEW YORK OXFORD
PARIS SAN DIEGO SAN FRANCISCO SINGAPORE SYDNEY TOKYO

Newnes
An imprint of Elsevier
Linacre House, Jordan Hill, Oxford OX2 8DP
200 Wheeler Road, Burlington, MA 01803

First published 1998
Second edition 2003

British Library Cataloguing in Publication Data
A catalogue record for this book is available from the British Library

ISBN 0 7506 59882

For information on all Newnes publications
visit our website at: www.newnespress.com

Typeset by Keyword Typesetting Services
Printed and bound in Great Britain by Biddles Ltd, *www.biddles.co.uk*

Contents

Preface

Sooner or later, most PC users find themselves confronted with hardware or software failure or the need to upgrade or optimize a system for some new application. *Newnes PC Troubleshooting Pocket Book* provides a concise and compact reference which describes in a clear and straightforward manner the principles and practice of fault finding and upgrading PCs.

This book is aimed at anyone who is involved with the installation, configuration, maintenance, upgrading, repair or support of PC systems. It also provides non-technical users with sufficient background information to diagnose basic faults and carry out simple modifications and repairs.

In computer troubleshooting, as with any field of endeavour, there are a number of short-cuts which can be instrumental in helping to avoid hours of frustration and costly effort. We have thus included a number of 'tips' which will help you avoid many of these pitfalls. Gleaned from a combined practical computing experience extending over 50 years, these snippets of information are the result of hard-won experience of the two co-authors and will hopefully save you hours of frustration!

To take into account the relentless advance in PC technology, this new edition has been considerably updated and extended and several completely new sections have been included on troubleshooting in the Windows environment and associated software.

Happy troubleshooting!
Michael Tooley, Howard Anderson

1 Introduction

PC troubleshooting covers a very wide variety of activities including diagnosing and correcting hardware faults and ensuring that systems are correctly configured for the applications which run on them. This chapter sets the scene for the rest of the book and explains the underlying principles of troubleshooting and fault finding.

This book makes very few assumptions about your previous experience and the level of underpinning knowledge which you may or may not have. You should at least be familiar with the basic constituents of a PC system: system unit, display, keyboard and mouse. In addition, you have probably had some experience of using Windows or perhaps even DOS.

Don't panic if you are a complete beginner to fault finding and repair. You can begin by tackling simple faults and slowly gain experience by moving on to progressively more difficult (and more challenging) faults. With very little experience you should be able to diagnose and rectify simple hardware problems, install a wide variety of upgrades, and optimize your system.

With more experience you will be able to tackle fault finding to 'replaceable unit' level. Examples of this could be diagnosing and replacing a faulty I/O card, a power supply, or a disk drive.

Fault finding to component level requires the greatest skill. It also requires an investment in specialized diagnostic equipment and tools. Nowadays, however, component level fault finding is often either impractical or uneconomic; you may require equipment available only to the specialist and it may be cheaper to replace a card or disk drive rather than spend several hours attempting to repair it.

1.1 A brief history of the PC

The original IBM PC was announced in 1981 and made its first appearance in 1982. The PC had an 8088 central processing unit and a mere 64K bytes of system board RAM. The basic RAM was, however, expandable to an almost unheard of total of 640K bytes. The original PC supported two 360K byte floppy disk drives, an 80 column × 25 line text display and up to 16 colours using a Colour Graphics Adapter (CGA).

The XT (eXtended Technology) version of the PC appeared in 1983. This machine provided users with a single 360K byte floppy drive and a 10M byte hard disk. This was later followed by AT (Advanced Technology) specification machines which were based on an 80286 microprocessor (rather than the 8088 used in its predecessors) together with 256K bytes of RAM fitted to the system board. The standard AT provided 1.2M bytes of floppy disk storage together with a 20M byte hard disk.

Not surprisingly, the standards set by IBM attracted much interest from other manufacturers, notable among whom were Compaq and Olivetti. These companies were not merely content to produce machines with an identical specification but went on to make their own significant improvements to IBM's basic specifications. Other manufacturers were happy to 'clone' the PC; indeed, one could be forgiven for thinking that the highest accolade that could be offered by the computer press was that a machine was 'IBM compatible'. This was because the PC was made from 'off-the-shelf' components. In a short time, almost identical machines became available at a fraction of the cost from a wide range of companies and became known as 'IBM compatible' PCs. These now dominate the market place. What were called IBM PCs are now just called PCs.

Since those early days, the IBM PC has become the 'de facto' standard for personal computing. Other manufacturers (such as Apple, Commodore and Atari) have produced systems with quite different specifications but none has been as phenomenally successful as IBM.

Pentium, AMD or Cyrix-based systems now provide performance specifications which would have been quite unheard of a decade ago and which have allowed software developers to produce an increasingly powerful and sophisticated range of products which will support multi-users on networked systems as well as single users running multiple tasks on stand-alone machines.

PCs are now produced by a very large number of well-known manufacturers. Machines are invariably produced to exacting specifications and you can be reasonably certain that the company will provide a good standard of after-sales service. Indeed, most reputable manufacturers will support their equipment for a number of years after it ceases to be part of a 'current product range'.

Many small companies assemble PC-compatible systems using individual components and boards imported from Far-East manufacturers. In many cases these systems offer performance specifications which rival those of well-known brands; however, the constituent parts may be of uncertain pedigree.

1.2 Conventions used in this book

The following conventions are used in this book:

1. Special keys and combinations of special keys are enclosed within angled braces and the simultaneous depression of two (or more keys) is indicated using a hyphen. Hence <SHIFT-F1> means 'press the shift key down and, while keeping it held down, press the F1 key'.
2. In addition, many of the special function keys (such as Control, Alternate, etc.) have been abbreviated. Thus <CTRL> refers to the Control key, <ALT> refers to the Alternate key and <DEL> refers to the Delete key. <CTRL-ALT-DEL> refers to the *simultaneous* depression of all three keys.
3. DOS commands and optional switches and parameters (where appropriate) have all been shown in upper case. In practice, DOS will invariably accept entries made in either upper or lower case. Thus, as far as DOS is concerned, **dir a:** is the same as **DIR A:**. For consistency we have used upper case but you may make entries in either upper or lower case, as desired.
4. Where several complete lines of text are to be entered (such as those required to create a batch file) each line should be terminated with the <ENTER> key. The ENTER key is also known as RETURN or CARRIAGE RETURN or even just CR. This name came from the pre-computing QWERTY typewriter keyboards on which PC keyboards are modelled.
5. Unless otherwise stated, addresses and data values are given in hexadecimal (base 16).
6. Finally, where several DOS commands are likely to be used together (e.g. within a batch file) or where we provide examples of output to a printer or a screen display, we have made use of a monospaced Courier font.
7. Internet URLs or URIs are given without the protocol as most browsers will assume the http protocol.

1.3 General approach to troubleshooting

Whatever your background it is important to develop a systematic approach to troubleshooting right from the start. This will help you to cope with obscure as well as routine faults.

> **TIP:** It is important to realize that most faults with PCs can be fixed by changing software or software settings, so avoid taking the 'lid' off until you have eliminated all software problems.
>
> In this context, 'hardware troubleshooting' in this book

includes fixing problems with the associated system *software* such as device drivers, etc. In modern practice, there is not very much scope in 'mending' the actual hardware, it is generally more expensive than buying a replacement.

TIP: Avoid 'upgrade-itis'. Only upgrade if you need the extra facilities or performance offered by the upgrade. There is no sense at all getting the latest version without knowing why. 'If it ain't broke, don't fix it.' For example, you may see that Microsoft 'will no longer support Windows 98'. This means nothing at all to the user, much help is still available, even from Microsoft. If Windows 98 works for your applications, on your machine, do not upgrade!

TIP: Beware the ash tray solution. The problem of a full ash tray in a car can easily be solved. Buy a new car. There are better ways to solve the problem! The ash tray solution is related to upgrade-itis and is encouraged by manufacturers. They will tell you that 'it is not compatible but if you buy the latest model, all your problems will be fixed'. In many people's experience, this approach often leads to other problems.

1.4 Where to start

It is perhaps worth saying that a system which appears to be totally dead can be a much easier prospect than one which displays an intermittent fault.

Start at the beginning and move progressively towards the end. This sounds obvious but many would-be troubleshooters ignore this advice and jump in at a later stage. By so doing, they often make erroneous assumptions and all too often ignore some crucial piece of information.

1.5 What to ask

If you are troubleshooting someone else's system you may be presented with a box and no information other than 'it doesn't work'. It has to be said that the average user is remarkably inadequate when it comes to describing faults on items of technical equipment. Furthermore he/she rarely connects the circumstances which lead up to equipment failure with the actual appearance of the fault. For example, a PC which has been relocated to a shelf over a radiator is bound to be a candidate for a very early death.

If you do have to deal with non-technical users it is well worth producing your own checklist of questions. To help you, the questions that we regularly use are as follows:

1. Has the fault just appeared or has it got progressively worse?
2. Is the fault present all the time?
3. If the fault is intermittent, under what circumstances does the fault appear?
4. Did the system work satisfactorily before? If not, in what way were you dissatisfied with its performance?
5. Has the configuration of the system changed in any way? If so, how has it changed?
6. What action (if any) have you taken to rectify the fault?
7. How did you first become aware of the fault?
8. Did you hear, see, or smell anything when the system failed?
9. What was actually happening when the system crashed?

In addition, you may wish to ask supplementary questions or make a few simple suggestions such as:

10. Have you checked the power to the system?
11. Is the printer on-line and is it loaded with paper?
12. Is the network 'up and running'?

In judging what reliance to place on the user's responses, it usually helps to make some assessment of the level of the user's technical expertise. You can do this by asking a few simple (but non-technical) questions and noting what comes back. Try something along the following lines:

13. How long have you been using the system?
14. Is this the first PC that you have used/owned?
15. How confident do you feel when you use the system?

In any event, it is important to have some empathy with the user and ensure that they do not feel insulted by your questions. A user who feels ignorant or threatened may often consciously or subconsciously withhold information. After all, the secretary who spills a cup of coffee over a keyboard is unlikely to admit to it within the boss's hearing.

1.6 Categorizing faults

It helps to divide faults into the following categories: hardware faults, software faults and configuration problems. This book is organized on this basis.

1.6.1 Suggested causes of trouble

The PC did work, now it gives trouble. No known hardware or software changes. It could be (in no particular order):

- Hardware failure.
- Software licence has run out.
- A virus has 'struck'.
- Run out of disk space.
- CMOS battery failure.
- Fuse blown (in power lead, inside power supply or on an internal card).
- Lead pulled out or damaged (*very* common!).
- Chip creep. This is where the chip slowly comes out of its socket due to the heating/cooling cycle when you turn the machine on/off each day. Press all socketed chips back into their sockets.
- Overheating.
- Unknown to you, 'helpful' people or children have made changes.
- Someone has not told you or does not understand they have changed the system, e.g. 'I only ran a CD from a PC games magazine...!'.
- Toddler has posted toys in any or all of the spaces/holes in the PC.
- Mobile phone interference.
- Strong magnetic field. (Causes odd colours on monitor, usual culprit, Hi-Fi speakers).
- PC was dropped.
- The layer of dust that forms on PC boards has become damp.

The PC did work, hardware has been changed. It could be (in no particular order):

- New hardware is faulty.
- Driver software for new hardware is faulty.
- New driver software is conflicting with other software.
- New hardware is conflicting with other hardware devices.
- New driver software has overwritten older but working software.
- New driver software is not correctly set up or configured.

The PC did work, software has been changed. It could be (in no particular order):

- New software is conflicting with other software.
- The registry has become corrupted.
- New software is faulty.
- New software has overwritten older but working software or software components.
- New software is not compatible with current operating system.
- New software is not correctly set up or configured.

Some of the above can be fixed without trouble. Detailed technical 'fixes' or updated drivers are usually found on the makers' website.

1.7 Hardware faults

Hardware faults are generally attributable to component malfunction or component failure. Electronic components do not generally wear out with age but they become less reliable at the end of their normal service life. It is very important to realize that component reliability is greatly reduced when components are operated at, or near, their maximum ratings. As an example, a capacitor rated at 25 V and operated at 10 V at a temperature of 20°C will exhibit a mean-time-to-failure (MTTF) of around 200 000 hours. When operated at 40°C with 20 V applied, however, its MTTF will be reduced by a factor of 10 to about 20 000 hours.

> **TIP:** The mean-time-to-failure (MTTF) of a system can be greatly extended by simply keeping it cool. Always ensure that your PC is kept out of direct sunlight and away from other heat producing sources (such as radiators). Ventilation slots should be kept clear of obstructions and there must be adequate air flow all around the system enclosure. For this reason it is important to avoid placing tower systems under desks, in corners, or sandwiched between shelves.

1.7.1 Hardware fault, what to do

If you think you have a hardware fault, the following stages are typical:

1. Perform functional tests and observations. If the fault has been reported by someone else, it is important to obtain all relevant information and not make any assumptions which may lead you along a blind alley.
2. Eliminate functional parts of the system from your investigation.
3. Isolate the problem to a particular area of the system. This will often involve associating the fault with one or more of the following:
   (a) power supply (including mains cable and fuse)
   (b) system motherboard (includes CPU, ROM and RAM)
   (c) graphics adapter (includes video RAM)
   (d) disk adapter (includes disk controller)
   (e) other I/O adapter cards (e.g. serial communications cards, modem cards, USB devices, SCSI devices, etc.)
   (f) floppy disk drive (including disk drive cables and connectors)

(g) hard disk drive (including disk drive cables and connectors)
(h) keyboard and mouse
(i) display adapter
(j) monitor
(k) external hardware (such as a printer sharer or external drive)
(l) communications or network problems

4. Disassemble (as necessary) and investigate individual components and subsystems (e.g. carry out RAM diagnostics, gain access to system board, remove suspect RAM).
5. Identify and replace faulty components (e.g. check RAM and replace with functional component).
6. Perform appropriate functional tests (e.g. rerun RAM diagnostics, check memory is fully operational).
7. Reassemble system and, if appropriate, 'soak test' or 'burn in' for an appropriate period.

TIP: If you have more than one system available, items such as the system unit, display, keyboard, and external cables can all be checked (and eliminated from further investigation) without having to remove or dismantle anything. Simply disconnect the suspect part and substitute the equivalent part from an identical or compatible system which is known to be functional.

1.8 Software faults

Software faults can arise from a number of causes including defective coding, corrupted data, viruses, 'software bombs' and 'Trojan horses'. Software faults attributable to defective coding can be minimized by good design practice and comprehensive software testing before a product is released. Unfortunately, this doesn't always happen. Furthermore, modern software is extremely complex and 'bugs' can often appear in 'finished' products due to quite unforeseen circumstances (such as changes in operating system code). Most reputable software houses respond favourably to reports from users and offer software upgrades, 'bug fixes' and 'workarounds' which can often be instrumental in overcoming most problems. The moral to this is that if you don't get satisfactory service from your software distributor/supplier you should tell all your friends and take your business elsewhere.

In recent years, computer viruses have become an increasing nuisance. A persistent virus can be extremely problematic and, in severe cases, can result in total loss of your precious data. You can avoid this trauma by adhering to a strict code of practice and by investing in a proprietary anti-virus package.

1.9 Configuration problems

Configuration problems exist when both hardware and software are operating correctly but neither has been optimized for use *with the other*. Incorrectly configured systems may operate slower or provide significantly reduced functionality when compared with their fully optimized counterparts. Unfortunately, there is a 'grey' area in which it is hard to decide upon whether a system has been correctly optimized as different software packages may require quite different configurations.

1.10 Burn-in

Any reputable manufacturer or distributor will check and 'burn in' (or 'soak test') a system prior to despatching it to the end user. This means running the system for several hours in an environment which simulates the range of operational conditions in which the system in question is likely to encounter.

'Burn-in' can be instrumental in detecting components that may quickly fail either due to defective manufacture or to incorrect specification. In the case of a PC, 'burn-in' should continuously exercise *all* parts of the system, including floppy and hard disk drives.

TIP: It is always wise to 'soak test' a system following any troubleshooting activity particularly if it involves the replacement of an item of hardware.

1.11 What is fitted in your PC?

There is some very good (free) software available that will tell you just what hardware and software is fitted in your machine.

See system info with belarc.com/
or
WCPUID obtainable from www.h-oda.com

You could look up this information using the Windows Control Panel 'System' icon but the above software is more comprehensive and is easier to use.

1.12 General points

Always perform one upgrade at a time. Windows often requires a reboot to complete an installation. Experience shows that when all is finished it is best to shut down, turn off all the power and restart before installing anything else. Windows 'restart' is never quite the same.

Do not be surprised if an apparently unrelated item interacts with the item you are working on. As an example, a flatbed scanner in use on the author's machine is fitted with an instant scan button. This can be enabled or disabled in the scanner settings. If it is enabled, problems occur with the printer. Nothing is 'broken', they simply conflict, even though the printer and scanner have nothing in common. They are on different buses – the scanner is a SCSI device, the printer is on the standard printer port.

Some people like to increase the speed of their PC by overclocking. Unless you know what you are doing, don't.
See www.overclockers.com.au/techstuff/

Read the guarantees carefully. Some are 'return to base, parts only' so may actually cost you more than buying a new part.

A very common response from help desks is to say 'RTFM'. This means 'Read The Flaming Manual'. Enough said.

Some software is difficult to uninstall, parts of it linger in the machine. If it causes no trouble then is it fine to leave it alone but sometimes third party uninstall software or registry tools are required.

2 The Internet

The net is a powerful tool and a huge resource for troubleshooting information. Some people even criticize the net as being too full of 'computer stuff'. Good news for a troubleshooter!

In some respects, the Internet does not exist. You cannot go to see it, it is not controlled by one governing body, it does not belong to anyone. Saying 'the Internet' is about as vague as saying 'the shops' or 'the roads', in both cases, all the shops or roads in the world. The Internet is not a 'thing', it is simply a large number of machines connected together sharing a common low level protocol, TCP/IP. Over the top of TCP/IP, use is made of a number of higher level protocols.

Sadly, some people seem to think the Internet is the world wide web. This is not the case, but the web (or more accurately the *http protocol*) is the most popular. Using the web, pages of information are accessed via a URL, a *Uniform Resource Locator*. URLs are made up as below:

http://www.w3.org/Addressing/URL/Overview.html

In this case, the *http://* is the protocol to use, the part that says *www.w3.org* is the *domain* or address of the server where the information is stored, the part *Addressing/URL/* is the file system directory (unfortunately, a directory is called a *folder* in Windows) where the file is stored and *Overview.html* is the actual file. The result of typing the URL into a browser is for an http request to be submitted to the web server at www.w3.org for the file Overview.html to be transmitted to your machine. If all is well, the web server will 'serve up' the web page to your browser. The browser software will then *render* the information on the screen according to the *html* content of the page. html is the HyperText Markup Language.

A protocol is simply 'a set of rules that define communication'. Other common Internet protocols are SMTP, which is used to control email, or FTP, which is used for file transfer.

You may see mention in some places of a URI instead of a URL.

The Internet standards are under constant review and the World Wide Web consortium is discussing the subject of URLs and trying to make them more universal. When implemented, they would then be called a Universal Resource Identifier or URI. In general use there is some confusion about what to use, URL or URI. If you would like the most up-to-date information, see http://www.w3.org/Addressing/URL/uri-spec.html

2.1 Internet, the main search engines

Since the Internet is not one 'thing', some means are required to find what you need. There is no central index so various people have devised ways to create indexes of Internet resources. These indexes are never complete so items will be 'on the Internet' but not indexed, they are 'there' but *search engines* will not find them. Something is 'on the Internet' when a machine with a TCP/IP connection is connected so it can reach other computers that in turn are connected to other computers, etc.

Some search engines have information organized by humans (also called a directory), others have information organized by computers. The second kind uses software (called variously a *spider*, *robot* or *crawler*) to look at each page on a site, extract the information and build an *index*. It is this index you search when using the search engine. Some people make a clear distinction between a directory and a search engine. Currently, the situation is not clear cut as many 'search engines' in fact use both methods and many share the same index!

The performance of a search engine depends critically on how well these indexes are built. It is also very important to remember that the whole business of search engines is in a state of constant change. Companies buy each other, change systems, indexes, etc. Some rely on other people's information. There is no such thing as a static search engine! The latest situation is usually available on www.searchenginewatch.com

Below are listed some of the common search engines:

Table 2.1 Common search engines

Google http://www.google.com	Google makes use of 'link analysis' as a way to rank pages. The more links to and from a page, the higher the ranking. They also provide search results to other search engines such as Yahoo.
AllTheWeb.com (FAST Search) http://www.alltheweb.com	One of the largest indexes of the web.
AltaVista http://www.altavista.com	One of the oldest crawler-based search engines on the web, it also has a large index of web pages and a wide range of searching commands. Many users now have to pay to be listed, limiting the usefulness of this engine.
AOL Search http://search.aol.com/	Uses the index from Open Directory and Inktomi and offers a different service to members and non-members.

Ask Jeeves http://www.askjeeves.com	Ask Jeeves is a human-powered search engine that introduced the idea of plain language search strings.
Direct Hit http://www.directhit.com	Direct Hit uses its own 'popularity engine' that depends on how many times a site is viewed to judge its ranking. This idea is not always successful as the less popular sites do not get a chance to rise, so popular ones remain popular. Direct Hit is owned by Ask Jeeves.
HotBot http://www.hotbot.com	Much of the time, HotBot's results come Direct Hit but other results come from Inktomi. Hotbot is owned by Lycos.
Inktomi http://www.inktomi.com	You cannot query the Inktomi index itself, it is only available through Inktomi's partners. Some 'search engines' simply relay what is found in the Inktomi index.
LookSmart http://www.looksmart.com	LookSmart is a human-compiled directory of websites but when a search fails, further results are provided by Inktomi.
Lycos http://www.lycos.com	Lycos uses a human developed directory similar to Yahoo and its main results come from AllTheWeb.com and Open Directory.
MSN Search http://search.msn.com	Microsoft's MSN is powered by LookSmart with other results from Inktomi and Direct Hit.
Netscape Search http://search.netscape.com	Netscape Search's results are from Open Directory and Netscape's own index. Other results are Google.
Open Directory http://dmoz.org/	Open Directory uses an index built by volunteers. It is owned by Netscape (who are owned by AOL).
Teoma http://www.teoma.com	A new search engine, launched in April 2002 that claims to be better than Google.
Yahoo http://www.yahoo.com	Yahoo is a human-compiled index but that uses information from Google.

2.2 Searching the web

Most people use search engines to find what they want on the net. Many have a favourite search engine on the grounds that it gives them what they want. The best advice is: *don't have a favourite*. You should realize that there are different kinds of search engine and each will (may!) find what you want depending on what that is. It is not unreasonable to use six or more search engines in a particular search.

This difference is not so much in the subject area, it is more on how you look and exactly what you want. It is important to remember what was outlined above: search engine indexes, the things that are actually searched, are either built by humans or by 'spider' software. Humans are good at subjects, software is better at indexing words. Yahoo at www.yahoo.com uses indexes built by people, so when someone submits a web address to Yahoo, a real live person decides on where to put it in the index. If you use www.go.com, you will be searching through an index created by software that looks for keywords in the webpages – it 'spiders'. Many search engines now use multiple indexes and some even use indexes of different types but the fact remains that different engines will give different results from the same search. *Use more than one.*

If you search for the words 'History' and 'Computer', you could get references to a file that contains the string 'he was playing with his toy computer during the history lesson' simply because it contains the correct words. If you had used a human categorized search engine, you are less likely to come across this problem. On the other hand, if keywords are what you want, a keyword search engine is better. For example, if you want information on colossus and Alan Turing, you would be better off with a keyword engine as neither colossus nor Alan Turing are 'subjects'. As a clear demonstration of the power of keyword search engines, try looking for a single line in a famous poem. Try looking for 'if you can keep your head when all about you' in both www.yahoo.com and www.go.com, making sure you put the string in quotes, you may be surprised at the different hits returned. (It is the first line in a poem called 'If' by Rudyard Kipling once voted 'the most popular poem in England'.) Even more impressive is looking for computer components. On a motherboard, a chip was found with just the number 'ms62256h' printed on it. The AltaVista Advanced search engine returned seven hits to data sheets about this SRAM chip, while www.google.com returned four. As another example, after a search for a piece of circuit board marked only with '5000532', Google, searched with just 5000532, gave a direct link to the maker. In this case, the board is a Gateway 256 MB 133/100/66 MHz 64 bit 4 clock 16 × 16 SDRAM DIMM!

Another consideration is search syntax. Use *Boolean* expressions. These contain logic symbols or statements like +, –, OR and AND. If you search for the words computer and history you should enter +computer +history, the + signs meaning that the word must be present. If you don't use the plus signs, some search engines will do a logical OR operation and search for either computer OR history. It is *well worth your trouble* looking at the search engine tutorial at www.brightplanet.com/deepcontent/index.asp, it will add much power to your searches. You know when you are getting good at searching the net when the number of hits you get from a search is less than 100. A search that gives a million hits is unlikely to be of much use. You can also use the – sign to mean do not include, e.g. +history +computer –mainframe to avoid the word mainframe.

Metasearch engines will search other engines for you and give good results. A problem with this approach is that they do not always pass on the right or full syntax to each client search engine, so carefully constructed Boolean searches do not always work.

If you do not have time (or the web address has changed since this was printed!) try these thoughts:

- Do *not* use CAPITAL letters in searches. Different engines use different rules about capital letters but lower case nearly always works.
- Put strings that contain spaces in quotes, e.g. 'if you can keep your head when all about you'.
- Use + and – signs *routinely*, some engines use OR and AND but most take + and –.
- Use the 'advanced' or 'power' searches. Some search engines only allow Boolean expressions in the advanced search page. You can then search for specific items such as 'stored program' AND electronic AND semiconductor AND silicon AND 'alan turing' AND 'blaise pascal'. On a recent trail with AltaVista advanced, this gave less than 40 hits and some interesting information (note no capital letters in the search string).
- Use plenty of keywords to narrow down a search, try +history +computer +microprocessor +intel instead of just +history +computer, the first returned 22 hits, the second over 2000 hits on a recent trial.
- Try putting the string in a different order. Some engines assume the first word is more important than the second so +history +computer may give different results from +computer +history.
- Use the * character as a wildcard, e.g. a search for comput* will find the words computer, computers, computing, computation, etc. Some search engines, notably www.google.com, do not allow wildcards.
- Avoid plurals such as 'computers' as it will miss the word 'computer'. It is better to use a wildcard like computer* which will find both.
- Use either site: or domain: to filter for different countries. In various engines, try +history +computer +pentium +site:uk. It should only give sites in the UK (but will miss those with a .com at the end).

Some subjects are hard to research and there are several reasons for this. Some commercial information is only available for a payment, for instance current business performance, stock prices, etc. are not generally available free of charge. Other information is swamped by commercial interests, e.g. try looking for +pentium +computer and you will find hundreds of companies trying to sell you their oh so cheap yet oh so fast computers. These are areas that will test your searching skills, often the solution is to use a Boolean search that includes a technical term not often found in sales literature. The search string +pentium +computer +silicon might typically return 90 mainly non-commercial hits whereas just +pentium +computer is likely to return over a million hits, many of them adverts.

Finally, do a search, using different engines, for the expression +search +engine. You will find much information that will aid your use of the fantastic resource called the Internet. Computing is one of the fastest changing subjects yet known and it is often difficult to keep up to date. One of the most profitable parts of your study time will be to polish your searching skills. As stated above, the latest situation is usually available on www.searchenginewatch.com.

2.3 An example of searching the Internet

Some subjects are hard to find as the words used in a search engine often lead to many different subjects.

The task here is to find specific information and to eliminate all the non-relevant information. There is no 'answer' as such, you either find what you want or not.

Suppose you wished to search for help with troubleshooting a monitor. You could search for *monitor has gone black.*

In a recent test of this *search string*, Google gave 380 000 hits, the first 'hit' gave:

> Help Desk FAQ's ... 11. My **monitor has gone black**. What do I do? Move the mouse or tap the Shift key on your keyboard. Sometimes monitors are set to . . .

i.e. possibly what was required, but the second hit gave:

> . . . Judith Matloff Staff writer of The Christian Science **Monitor**. . . . the shorter, cheaper route to Georgia's **Black** Sea port ... over the past century oil has **gone up** and . . .

i.e. nothing to do with computer monitors.

Lycos gave an even larger number of hits, 1 213 516, the first was:

> man with **black** hat Fast Forward » – . . . ad | advertise here man with **black** hat Wednesday, November . . . Fernandes Nomad, Standard Model, in black. This is my first solid-body . . . since I was a kid. But this one **has** a built-in amplifier/speaker . . . into the desert . . .' (Matt 4:1)It **has** become necessary to be away from . . .

i.e. nothing to do with broken monitors, the second was:-

> Monitor Repair Fast Forward » –. . . know the value of zd301. **Monitor has** been shelved for a while...P101702-6: IBM 6547-0BN **monitor has** a faulty HOT SEC5088 . . . Acer 54el FCC JVP7154E **monitor has** no display. Replaced . . . P102102-3: Compaq 462 **monitor has** no brightness control . . .

The point is, if search engines look for keywords, you do not get results by subject.

A results list of 1 213 516 or 380 000 hits is far too large, it is better to refine the search.

Putting the search string in quotes, e.g. 'monitor has gone black', gave ten hits from Google, all of which referred to monitor trouble; Lycos gave 11, all relevant.

In general, keyword searches are better provided you think about the keywords you search for.

2.4 Can you trust the answers you find?

That depends on who published the information. The problem is no different when considering the Internet, magazines or books.

If the site owner is a private individual, the information may be correct but it is not likely to have been checked with great rigour. If it tallies with similar sites, the chances are good that it is correct, but on the other hand, who copied who?

If it is an official site, the information will generally be an accurate reflection of the opinion of the site owners. Manufacturers will tell you their products are wonderful, governments will tell you they serve you to the highest standards. Make your own judgement.

Academic sites often contain information from a more 'free-thinking' group of people so will show a wide range of opinion. Specific scientific, technological or historical information is usually accurate, political views may be very varied. Again, make your own judgement.

3 Microcomputer fundamentals

An understanding of the basic operation of a microcomputer system is an essential first step to getting the best out of your PC. This chapter provides the basic underpinning knowledge required to carry out successful upgrading and troubleshooting.

The chapter begins by describing the basic components of a microcomputer and how data is represented within it. The chapter includes a quick tour of a system with particular reference to the central processing unit (CPU), memory (ROM and RAM), and the means of input and output. The chapter concludes with a brief introduction to some of the facilities provided by an operating system.

3.1 Microcomputer basics

The basic components of a microcomputer system are:

- A central processing unit (CPU).
- A memory, comprising both 'read/write' and 'read-only' devices (commonly called RAM and ROM respectively).
- A mass storage device for programs and/or data (e.g. a floppy and/or hard disk drive).
- A means of providing user input and output (via a keyboard and display interface).
- Interface circuits for external input and output (I/O). These circuits (commonly called 'ports') simplify the connection of peripheral devices such as printers, modems, mice, and joysticks.

In a microcomputer (as distinct from a mini or mainframe machine) the functions of the CPU are provided by a single VLSI microprocessor chip (e.g. an Intel 8086, 8088, 80286, 80386, 80486, or Pentium). The microprocessor is crucial to the overall performance of the system. Indeed, successive generations of PC are normally categorized by reference to the type of chip used. The 'original' PC used an 8088, AT systems are based on an 80286, '386 machines use an 80386, and so on.

Semiconductor devices are also used for the fast read/write and read-only memory. Strictly speaking, both types of memory permit 'random access' since any item of data can be retrieved with equal ease regardless of its actual location within the memory. Despite this, the term 'RAM' has become synonymous with semiconductor read/write memory. (VLSI means very large scale integration, i.e. a complex chip.)

The semiconductor ROM provides non-volatile storage for part of the operating system code (this 'BIOS' code remains intact when the power supply is disconnected). The semiconductor RAM provides storage for the remainder of the operating system code (the 'DOS'), applications programs and transient data (including that which corresponds to the screen display).

It is important to note that any program or data stored in RAM will be lost when the power supply is switched off or disconnected. The only exception to this is a small amount of 'CMOS memory' kept alive by means of a battery. This 'battery-backed' memory is used to retain important configuration data, such as the type of hard and floppy disk fitted to the system and the amount of RAM present.

> **TIP:** It is well worth noting down the contents of the CMOS memory to avoid the frustration of having to puzzle out the settings for your own particular system when the backup battery eventually fails and has to be replaced. To view the current CMOS configuration settings press the 'Del' key during the bootup sequence and enter the 'Setup' routine.

3.2 Catching the bus

The basic components of a simple microcomputer system, the CPU, RAM, ROM and I/O system, are linked together using a multiple-wire connecting system know as a 'bus' (see Figure 3.1). Three different buses are present (together with any specialized 'local' buses used for high-speed data transfer). The three main buses are:

- An 'address bus', used to specify memory locations.
- A 'data bus', on which data is transferred between devices.
- A 'control bus', which provides timing and control signals throughout the system.

NB PCs use a more complex architecture that will be shown in Chapter 5, 'System Architecture and Construction'.

3.3 Expanding the system

In the generalized system shown in Figure 3.1, we have included the keyboard, display and disk interface within the block marked 'I/O'. The IBM PC provides the user with somewhat greater flexibility by making the bus and power connections available at a number of 'expansion connectors'. The connectors permit the use of 'adapter cards' (see Figure 3.2). These adapters allow the system to be configured for

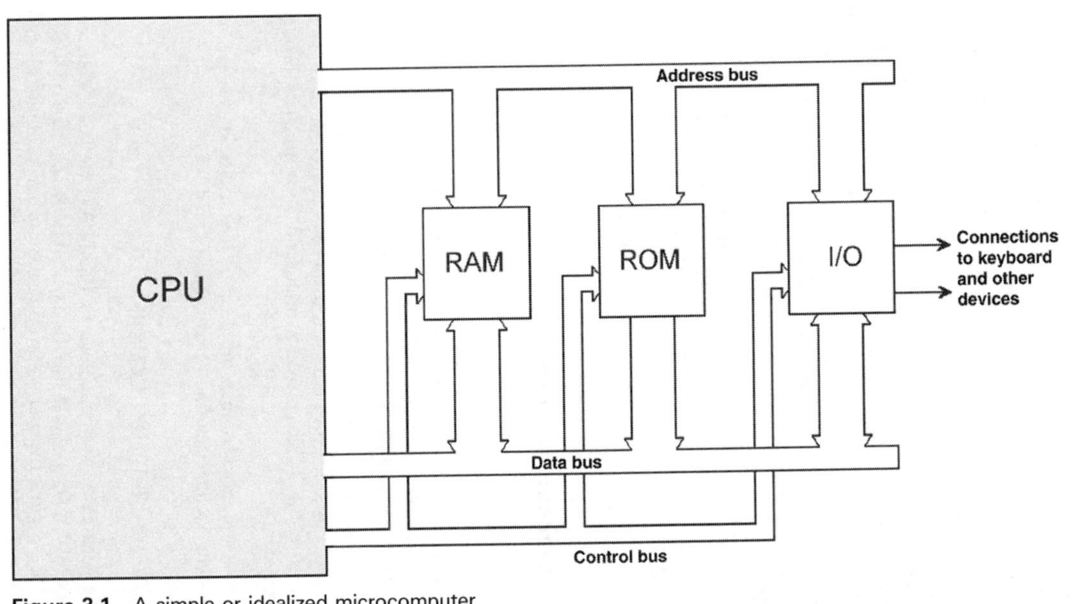

Figure 3.1 A simple or idealized microcomputer

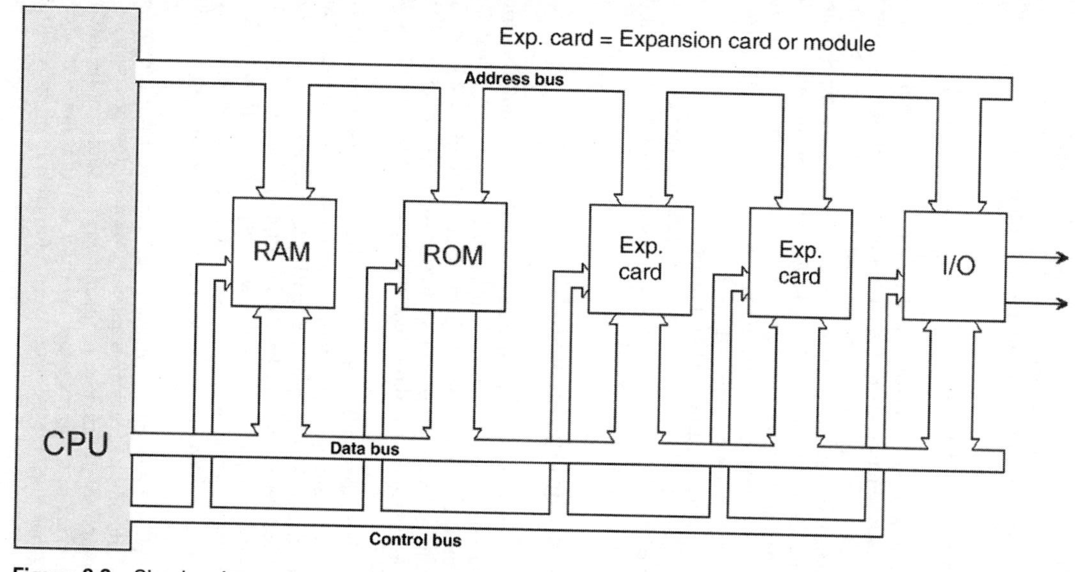

Figure 3.2 Simple microcomputer with expansion cards on common bus

different types of display, mass storage device, etc. Commonly available expansion cards include floppy and hard disk adapters, expansion memory cards, games (joystick) adapters, sound and video cards, internal modems, CD-ROM cards, and additional serial/parallel ports.

3.4 Clocks and timing

To distinguish valid data from the transient and indeterminate states that occur when data is changing, all bus data transfers must occur at known times within a regular cycle of 'reading' and 'writing'. Therefore the movement of data around a microcomputer system is synchronized using a master 'system clock'. This signal is the basic heartbeat of the system; the faster the clock frequency the smaller the time taken to execute a single machine instruction. The clock is a series of logical '1s' and '0s' and has nothing to do with timekeepers!

The basic timing is generated by a quartz crystal. This device ensures that the clock signal is both highly accurate and extremely stable. On the original PC, the 'system clock' signal was obtained by dividing this fundamental output frequency by a factor of 4.

3.5 Interrupting the system

Another control signal of particular note is the 'interrupt'. Interrupts provide an efficient means of responding to the needs of external hardware, such as a keyboard or a modem connected to the serial port. The Intel family of processors provides interrupts which are both 'maskable' and 'non-maskable', i.e. those that can be turned off and those that cannot.

When a non-maskable interrupt input is asserted, the processor must suspend execution of the current instruction and respond immediately to the interrupt. In the case of a maskable interrupt, the processor's response will depend upon whether interrupts are currently enabled or disabled (when enabled, the CPU will suspend its current task and carry out the requisite interrupt service routine). The response to interrupts can be enabled or disabled by means of program instructions (EI and DI respectively).

In practice, interrupt signals may be generated from a number of sources and since each will require its own customized response, a mechanism must be provided for identifying the source of the interrupt and vectoring to the appropriate interrupt service routine. In order to assist in this task, the PC uses a programmable interrupt controller, part of what has become known as the 'chip set'. A further type of interrupt is generated by software. These 'software interrupts' provide an efficient means of accessing the operating system services.

Interrupts are used to achieve 'multitasking'. This is where the CPU is made to switch between tasks at high speed giving the appearance of running several tasks at the same time. This is an illusion as one CPU can only run one task at a time!

3.6 Data representation

The number of individual lines present within the address bus and data bus depends upon the particular microprocessor employed (see Table 3.1). Some processors (notably the old 80386SX, 80486SX, etc.) only have a 16-bit external data bus to permit the use of a lower-cost motherboard while still retaining software compatibility with their full bus width processors (such as the 80386DX, 80486DX, etc.).

Signals on all lines, no matter whether they are used for address, data, or control, can exist in only two basic states: logic 0 ('low') or logic 1 ('high'). Data and addresses are represented by binary numbers (a sequence of 1s and 0s) that appear respectively on the data and address bus.

The largest binary number that can appear on a 16-bit data bus corresponds to the condition when all 16 of the lines are at logic 1. Therefore the largest value of data that can be present on the bus at any instant of time is equivalent to the binary number 1111111111111111 (or 65 535). Similarly, the highest address that can appear on a 20-bit address bus is 11111111111111111111 (or 1 048 575).

Table 3.1 Crude indicators for Intel microprocessors

Intel chip	Date	MIPS	Width of data bus	Number of transistors
4004	1971	0.06	4	2 300
8008	1972	0.06	8	3 500
8080	1974	0.64	8	6 000
8085	1976	0.37	8	6 500
8086	1978	0.33	16	29 000
8088	1979	0.33	16	29 000
80286	1982	1.2	16	134 000
80386SX	1985	5.5	16	275 000
80486DX	1989	20	32	1 200 000
80486SX	1991	13	32	1 185 000
80486DX2	1992	41	32	1 200 000
80486DX4	1994	52	32	1 600 000
Pentium P5	1993	100	64	3 100 000
Pentium P54C	1994	150	64	3 200 000
Pentium MMX	1997	278	64	4 500 000
Pentium Pro	1995	337	64	5 500 000

3.7 Binary and hexadecimal

For convenience, the binary data present within a system is often converted to hexadecimal (base 16). This format is easier for mere humans to comprehend and offers the advantage over denary (base 10) in that it can be converted to and from binary with ease. Some numbers in binary, denary and hexadecimal are shown in Table 3.2. A single hexadecimal character (in the range 0 to F) is used to represent a group of four binary digits (bits).

A 'byte' of data comprises a group of eight bits. Thus a byte can be represented by just two hex characters. A group of 16 bits can be represented by four hex characters, 32 bits by eight hex characters, and so on.

> **TIP:** The value of a byte expressed in binary can be easily converted to hexadecimal by arranging the bits in groups of four and converting each nibble into hexadecimal using Table 3.2. Taking 10100011 as an example: 1010 = A and 0011 = 3 thus 0100011 can be represented by hex A3.

3.7.1 Data in memory

A byte of data can be stored at each address within the total memory space of a computer. Hence one byte can be stored at each of the 1 048 576 memory locations within a machine offering 1 Mbyte of RAM.

3.8 A quick tour of the system

To explain the operation of the microcomputer system shown in Figure 3.2 in greater detail, we shall examine each major system component individually. We shall start with the single most important component of the system, the CPU.

3.8.1 The CPU

The CPU forms the heart of any microcomputer and, consequently, its operation is crucial to the entire system. The primary function of the microprocessor is that of fetching, decoding and executing instructions resident in memory. As such, it must be able to transfer data from external memory into its own internal registers and vice versa. Furthermore, it must operate predictably, distinguishing, for example, between an operation contained within an instruction and any

Table 3.2 Example decimal, hex and binary values

Decimal	Hex	Binary	Decimal	Hex	Binary	Decimal	Hex	Binary
0	0	0	30	1E	11110	32	0	100000
1	1	1	40	28	101000	48	30	110000
2	2	10	50	32	110010	64	40	1000000
3	3	11	60	3C	111100	80	50	1010000
4	4	100	70	46	1000110	96	60	1100000
5	5	101	80	50	1010000	112	70	1110000
6	6	110	90	5A	1011010	128	80	10000000
7	7	111	100	64	1100100	144	90	10010000
8	8	1000	200	C8	11001000	160	A0	10100000
9	9	1001	300	12C	100101100	176	B0	10110000
10	A	1010	400	190	110010000	192	C0	11000000
11	B	1011	500	1F4	111110100	208	D0	11010000
12	C	1100	600	258	1001011000	224	E0	11100000
13	D	1101	700	2BC	1010111100	240	F0	11110000
14	E	1110	800	320	1100100000	256	100	100000000
15	F	1111	900	384	1110000100	1024	400	10000000000
16	10	10000	1000	3E8	1111101000	2048	800	100000000000
17	11	10001	2000	7D0	11111010000	3072	C00	110000000000
18	12	10010	4000	FA0	111110100000	3840	F00	111100000000
19	13	10011	8000	1F40	1111101000000	4095	FFF	111111111111
20	14	10100	16 000	3E80	11111010000000	65 535	FFFF	1111111111111111

accompanying addresses of read/write memory locations. In addition, various system housekeeping tasks need to be performed including responding to interrupts from external devices.

The main parts of a microprocessor are:

- Registers for temporary storage of addresses and data (MAR, AC and SDR in Figure 3.3).
- An 'arithmetic logic unit' (ALU) that performs arithmetic and logic operations.
- A means of controlling and timing operations within the system, the control unit (CU).

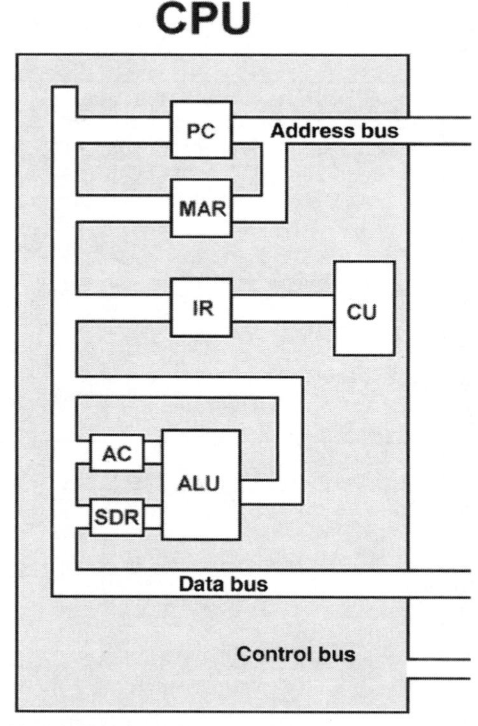

Figure 3.3 Model CPU (real CPUs are more complex but share the same basic idea)

It is important to remember that the CPU and RAM are separate so instructions or data stored in RAM must be 'fetched' from RAM before they can be 'executed'. The speed of this fetch–execute cycle is critical to the performance of the CPU.

The majority of operations performed by a microprocessor involve the movement of data. The program code, a set of instructions stored in memory, must itself be fetched from memory, piece by piece, prior to execution. The microprocessor thus performs a continuous sequence of instruction fetch and execute cycles. The act of fetching an instruction code (or operand or data value) from memory involves a read operation while the act of moving data from the microprocessor to a memory location involves a write operation.

Microprocessors determine the source of data (when it is being read) and the destination of data (when it is being written) by placing a unique address on the address bus. The address at which the data is to be placed (during a write operation) or from which it is to be fetched (during a read operation) can either constitute part of the memory of the system (in which case it may be within ROM or RAM) or it can be considered to be associated with input/output (I/O).

Since the data bus is connected to a number of VLSI devices, an essential requirement of such chips (e.g. ROM or RAM) is that their data outputs should be capable of being isolated from the bus whenever necessary. These VLSI devices are fitted with select or enable inputs which are driven by address decoding logic (not shown in Figures 3.2 or 3.3). This logic ensures that ROM, RAM and I/O devices never simultaneously attempt to place data on the bus!

The inputs of the address decoding logic are derived from one, or more, of the address bus lines. The address decoder effectively divides the available memory into blocks corresponding to a particular function (ROM, RAM, I/O, etc.). Hence, where the processor is reading and writing to RAM, for example, the address decoding logic will ensure that only the RAM is selected while the ROM and I/O remain isolated from the data bus.

Within the CPU, data is stored in several 'registers'. Registers themselves can be thought of as a simple pigeon-hole arrangement that can store as many bits as there are holes available. Generally, these devices can store groups of 16 or 32 bits. Additionally, some registers may be configured as either one register of 16 bits or two registers of 32 bits.

Some microprocessor registers are accessible to the programmer whereas others are used by the microprocessor itself. Registers may be classified as either 'general purpose' or 'dedicated'. In the latter case a particular function is associated with the register, such as holding the result of an operation or signalling the result of a comparison.

The ALU can perform arithmetic operations (addition and subtraction) and logic (complementation, logical AND, logical OR, etc.). The ALU operates on two inputs and it provides one output. In addition, the ALU status is preserved in the 'flag register' so that, for example, an overflow, zero or negative result can be detected.

The control unit is responsible for the movement of data within the CPU and the management of control signals, both internal and external. The control unit asserts the requisite signals to read or write data as appropriate to the current instruction.

3.8.2 Parallel input and output

The transfer of data within a microprocessor system involves moving groups of 8, 16, 32 or 64 bits using the bus architecture described earlier. Consequently it is a relatively simple matter to transfer data into and out of the system in parallel form. This process is further simplified by using a dedicated I/O device. This device provides registers for the temporary storage of data that not only 'buffer' the data but also provide a degree of electrical isolation from the system data bus.

3.8.3 Serial input and output

Parallel data transfer is primarily suited to high-speed operation over relatively short distances, a typical example being the linking of a microcomputer to an adjacent printer. There are, however, some applications in which parallel data transfer is inappropriate, for example by means of telephone lines. In such cases data must be sent serially (one bit after another) rather than in parallel form.

To transmit data in serial form, the parallel data from the microprocessor must be reorganized into a stream of bits. This task is greatly simplified by using an LSI interface device that contains a shift register which is loaded with parallel data from the data bus. This data is then read out as a serial bit stream by successive shifting. The reverse process, serial-to-parallel conversion, also uses a shift register. Here data is loaded in serial form, each bit shifting further into the register until it becomes full. Data is then placed simultaneously on the parallel output lines.

3.9 Operating systems

Many of the functions of an operating system (like those associated with disk filing) are obvious. Others, however, are so closely related to the machine's hardware that the average user remains blissfully unaware of them. This, of course, is as it should be. As far as most end users of computer systems are concerned, the operating system provides an environment from which it is possible to launch and run applications programs and to carry out elementary maintenance of disk files. Here, the operating system is perhaps better described as a 'computer resource manager'.

The operating system provides an essential bridge between the user's application programs and the system hardware. In order to provide a standardized environment (which will cater for a variety of different hardware configurations) and ensure a high degree of software portability, part of the operating system is hardware independent. The hardware dependent remainder (the 'BIOS') provides the individual low-level routines required by the machine in question.

A well-behaved applications program will interact with the hardware independent routines. These, in turn, will interact with the lower-level hardware dependent (BIOS) routines. Figure 3.4 illustrates this important point. This design is called 'layered', each set of function is built into a 'layer' in the system.

The user
↓
Application software, e.g.
wordprocessor
↓
Operating system user
interface (Windows, Linux)
↓
Hardware independent part of
operating system
↓
BIOS
(basic input output system)
↓
System hardware

Figure 3.4 Layered architecture of an operating system

The operating system also provides the user with a number of utility programs which can be used for housekeeping tasks such as disk formatting, disk copying, etc.

In order to provide a means of interaction with the user, the operating system incorporates either a GUI (graphic user interface) used with a pointer such as a mouse and a keyboard or a shell program used via keyboard entered commands and on-screen prompts and messages. A shell program is another name for a command line interpreter or CLI. Examples of a shell program are COMMAND.COM provided within

MS-DOS or *bsh*, the Berkeley shell available in the unix operating system.

In order to optimize the use of the available memory, most modern operating systems employ memory management techniques which allocate memory to transient programs and then release the memory when the program is terminated. This can cause problems when software fails to release the memory after use. If a program fails in this respect, it is called 'memory leakage' and is a common cause of seemingly intermittent problems. If a program such as Word 97 is used to open and close many documents in a session, sometimes the system reports that there is not sufficient memory to run the application. After restarting the machine, all seems well. This is an example of software that does not efficiently release memory used when it is no longer required, i.e. it remains 'allocated' by the memory management system so it appears there is not enough memory.

3.10 Dismantling a system

The procedure for dismantling a system depends upon the type of case or enclosure.

> **TIP:** Some cases have very sharp edges inside, especially those made of steel. Take great care not to slide any part of your body along these edges as painful injuries may result.

- Exit from any program that may be running.
- Shutdown the operating system.
- Switch the system unit power off.
- Switch off at the mains outlet and disconnect the mains power lead.
- Switch off and disconnect any peripherals that may be attached (including keyboard, mouse, printer, etc.).
- Disconnect the display power lead and video signal cable from the rear of the system unit. Remove the display and place safely on one side.
- Remove the cover retaining screws from the rear of the system unit.
- Carefully slide the system unit cover away. This can be awkward with some designs, some cases have internal 'hooks' that keep one side of the case in place. When the cover will slide no further, remove it and set aside.
- You will now have access to the system board, power supply, disk drives and adapter cards.

> **TIP:** Many people find that a battery powered electric screwdriver is very useful when working with PCs.

TIP: An egg box or similar container makes an excellent receptacle for screws and small parts when you are dismantling a system.

TIP: When a system uses a number of screws of similar size but of differing length, it is important to note the location of each screw so that it can be replaced correctly during re-assembly. A water-based felt-tip pen can be used to mark the screw sizes on the case.

TIP: If you are building or assembling your own system, always start with the largest size enclosure. This will provide you with plenty of scope for expansion and allow you to upgrade more easily.

3.10.1 Reassembly

System unit reassembly is usually the reverse of disassembly. It is, of course, essential to check the orientation of any non-polarized cables and connectors and also to ensure that screws have been correctly located and tightened. Under no circumstances should there be any loose connectors, components, or screws left inside a system unit!

3.11 Safety first!

The voltages found in mains-operated PC equipment can be lethal. However, high voltages are normally restricted to the power supply and display. The lower voltages present on the system board, disk drives and adapter cards are perfectly safe to *you* but static on your hands can damage the PC circuits.

When working inside the power supply or the display it is *essential* to avoid contact with any metal parts or components which may be at a high voltage. This includes all mains wiring, fuses and switches, as well as many of the components associated with the high-voltage a.c. and d.c. circuits in the power supply.

It is always essential to switch off and allow the capacitors to discharge before attempting to remove or replace components. Occasionally, you may have to test and/or make adjustments on 'live' circuits. In this event you can avoid electric shock hazards by only using tools which are properly insulated, and by using test leads fitted with insulated test prods. It is usually better to replace a suspected defective power supply with a new one and also to replace a display with another that's known to be working rather than attempt board-level or component-level servicing.

TIP: Another sensible precaution when making high-voltage measurements and adjustments is that of only working with one hand (you should keep the other one safely behind your back or in a pocket). This simple practice will ensure that you never place yourself in a position where electric current will pass from one hand and arm to the other via your heart. In such circumstances an electric shock could be fatal. If you are not certain you know what you are doing, ask!

3.12 Static hazards

Many of the devices used in modern PC equipment are susceptible to damage from stray electrostatic charges. Static is, however, not a problem provided you observe the following simple rules:

1. Ensure that your test equipment is properly earthed.
2. Preserve the anti-static wrapping supplied with boards and components and ensure that it is used for storage and also whenever boards or components have to be returned to suppliers.
3. Invest in an anti-static mat, grounding wire and wrist strap and use them whenever you remove or replace components fitted to a PCB.
4. Check your workshop or work area for potential static hazards (e.g. carpets manufactured from man-made fibres, clothing made from synthetic materials, etc.).

TIP: When working within the system unit make sure that you ground yourself by touching any grounded metal part (e.g. the case of the power supply) before removing or replacing any parts or adapter cards.

TIP: When components are mounted on a PCB there are plenty of paths which will allow static charges to drain safely away. Hence you are unlikely to damage components by touching them when they are in their correct locations on a PCB.

3.13 Cooling

All PC systems produce heat and some systems produce more heat than others. Adequate ventilation is thus an essential consideration and fans are included within the system unit to ensure that there is adequate air flow. Furthermore, internal air flow must be arranged so that it is unrestricted as modern processors and support chips run at high temperatures. These devices are much more prone to failure when they run excessively hot than when they run cool or merely warm.

If the system unit fan fails to operate (and it is not thermostatically controlled) check the supply to it. If necessary replace the fan. If the unit runs slowly or intermittently it should similarly be replaced. Some motherboards allow CPU temperature monitoring and control.

See these links for details of cooling fans:

www.directron.com/cases—case-fans.html
www.thermaltake.com/support/CoolingGuide.htm

3.13.1 Software control of cooling

Some modern motherboards have temperature sensors that can be used to give warning of an excessive temperature. See an example on:

www.cpetc.com/products/motherboards/ae25r.html

> **TIP:** CPUs produce a considerable amount of heat and often run at an excessive temperature. You can significantly improve the reliability of the processor (and greatly extend its working life) by fitting an effective cooling fan.

> **TIP:** When fitting expansion cards and positioning internal ribbon cables, give some thought to the air flow within the system unit. In particular, it is worth trying to maximize the space between adapter cards (rather than have them sandwiched close together). You should also ensure that the ribbon cables do not impede the flow of air around the motherboard and adapter cards.

3.13.2 Problems related to cooling fans

Noise

Fans that are old, and hence have worn bearings, can emit an irritating noise. Fans are very cheap so should be replaced.

Noise and dust

Fans that have been in use for some time will be dirty. The air they move over the components is not filtered so some of the dust sticks to the blades of the fan. This often causes the fan to go out of balance and make more noise. A good cleaning with a small paint brush will fix it.

Dust

In a typical PC, the fan will draw air in through the CD-ROM drive, floppy drive or whatever has a hole in it at the front. This air generally escapes at the back, leaving the dust that it contained clogging the spaces where it has flowed and usually leaving a layer of dust on the motherboard. If this layer of dust becomes damp, it will conduct a small amount of electric current, causing the PC to fail in damp weather. It may come as a surprise to see how much dirt accumulates in just a year or two of operation.

4 System architecture and construction

4.1 PC architecture

The term 'PC' now applies to such a wide range of equipment that it is difficult to pin down the essential ingredients of such a machine. However, at the risk of oversimplifying matters, a 'PC' need only satisfy two essential criteria:

- Be based upon an Intel, AMD, Cyrix or similar processor.
- Be able to support the PC operating systems such as Linux, DOS or the Microsoft Windows range.

The generic PC, whether a 'desktop' or 'tower' system, comprises three units: *system unit, keyboard* and *display*.

A typical system unit contains a number of items including:

- The system board or 'motherboard' to which is attached a number of memory modules and adapter cards.
- The power supply.
- One or more floppy drive(s).
- One or more hard drives(s).
- One or more CD-ROM/DVD drive(s).

Fortunately, all of these units are fairly easy to spot and easy to recognize.

4.2 Modern system board layouts

Figure 4.1 shows a typical PC board layout. This is somewhat more complex than Figure 3.1 but it still has memory and CPU separate, so will follow the fetch–execute sequence.

- The CPU is the central processing unit, i.e. the Pentium or AMD chip.
- The system RAM is the main memory typically fitted as DIMMs or similar.
- The FSB or front side bus is the main communication to/from the CPU. The speed of the FSB is critical to the performance of the whole machine.
- The PCI or peripheral component interconnect bus is an Intel design to connect adapter cards to the main system.

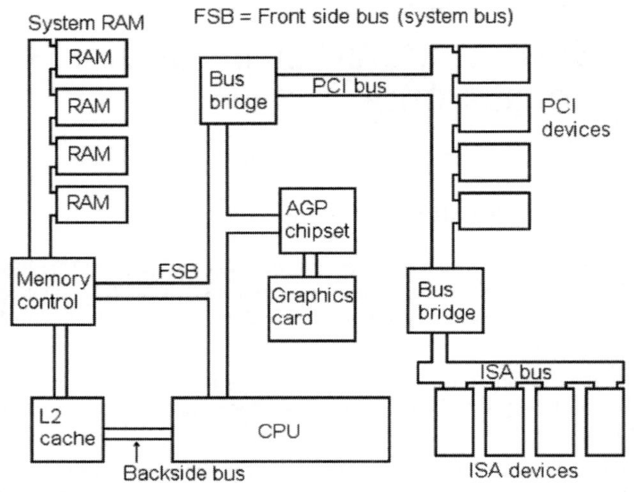

Figure 4.1 Typical PC board layout

- The ISA or industry standard architecture bus is also known as a legacy bus. This is the old, slow 8- or 16-bit bus fitted to original PCs. It has disappeared from some modern boards and it is only included on others to allow the fitting of older devices.
- The AGP or advanced graphics port is a high-speed link directly to/from the CPU to allow improved graphics performance. This is missing from some older boards.
- A bus bridge is simply a device that 'converts' the signals on one type of bus to another, it will handle speed and signalling differences.

There is another way of looking at a PC motherboard layout. Figure 4.2 shows essentially the same PC as in Figure 4.1 but in a more modern way. Older PCs were made from numerous chips, assembled to make the complete machine. Modern machines contain most functions in dedicated *chipsets* that are used to connect the CPU with the rest of the machine. This chipset is split into two main parts, the *north bridge* and the *south bridge* as in Figure 4.2. Intel now calls the south bridge the *communications I/O controller hub* (C-ICH or just ICH). The north bridge is now called the *graphics memory control hub* (GMCH).

The design and speed of the chipset plays a vital role in the overall performance of the machine.

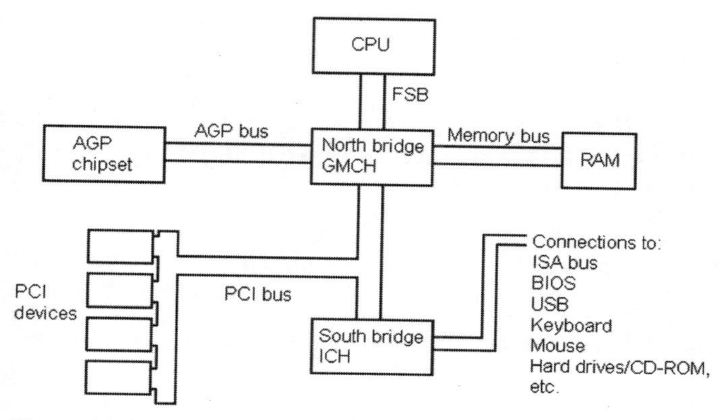

Figure 4.2 PC chipset

> **Historical note:** With PC specifications things may not always be what they seem. The original Pentium processor with its 64-bit data bus promised to offer PC users the advantages of 64-bit processing. In fact, Pentium architecture is based on two interconnected 32-bit '486-type processors. When the Pentium was first launched, it was sobering to find that the first generation of these much heralded chips could only just match the speed of the 'clock doubled' '486 chips that they were designed to replace (real benefits didn't materialize until the much faster Pentiums appeared). As far as memory is concerned and because of its 32-bit address bus, the Pentium is able to address exactly the same range as its predecessor. Not surprisingly, many people who rushed out to purchase the first Pentium-based systems were very disappointed with their performance – there must be a moral here somewhere!
>
> You should be very careful about manufacturers' claims and be especially wary of 'the numbers', fantastic clock speeds and amounts of RAM. The only real indicator is how fast the machine does the work you want it to perform.

4.3 Wiring and cabling

Internal wiring within a PC tends to take one of three forms:

- Power connections based on colour coded stranded wires (red, black, yellow, etc.)

- Ribbon cables (flat, multi-core wiring which is often grey or beige in colour).
- Signal wiring (miniature colour coded wires with stranded conductors) used to connect front panel indicators, switches, etc.

TIP: Ribbon cables invariably have a coloured stripe at one side which denotes the position of pin-1 on the connector. Since some connectors are 'non-polarized' (i.e. it is possible to make the connection the 'wrong way round') you should always carefully check that the stripe is aligned towards the '1' marked on the PCB. Making the connection the 'wrong way round' can sometimes have disastrous consequences.

4.3.1 Colour coding

The power supply wiring is invariably colour coded. The colour coding often obeys the following convention but different colours may be used to denote other power supply voltage rails and signals:

Red	+5 V	Used for main system +5 V supply rail
Yellow	+12 V	An ancillary supply rail used by disk drives, etc. to power motors
Black	0 V ground/common/	This variously named rail links all ground and chassis points and also acts as the negative 'return' connection for the +5 V and +12 V rails

4.4 Replacing the CPU

The processor chip (regardless of type) is invariably fitted in a socket or a slot and this makes removal and replacement quite straightforward provided that you take reasonable precautions.

The following describes the stages in removing and replacing a CPU chip:

1. Switch 'off', disconnect from the supply and gain access to the system board.
2. Ensure that you observe the safety and static precautions at all times. Have some anti-static packing available to receive the CPU when it has been removed.
3. Locate the CPU and ensure that there is sufficient room to work all around it (you may have to move ribbon cables or adapter

cards to gain sufficient clearance to use the extraction and insertion tools).

4. Depending on the design of the socket/slot, release the catch that holds the CPU in place.
5. Immediately deposit the chip in an anti-static container (do not touch any of the pins).
6. Pick up the replacement chip from its anti-static packing. Position the insertion chip over the socket and ensure that it is correctly orientated.
7. Reassemble the system (replacing any adapter cards and cables that may have been removed in order to gain access or clearance around the CPU). Reconnect the system and test.

4.5 Upgrading the CPU

A relentless increase in the power of the CPU makes this particular component a prime candidate for upgrading a system in order to keep pace with improvements in technology. Figures 4.3 and 4.4 show how the power of the Intel family of processors has increased over the last few decades.

Moore's law says that the number of transistors used in microprocessors will double every 18 months. The progress seems to correlate well with this 'law'.

Although Moore's law refers to the number of transistors in an integrated circuit, the clock speed of Intel processors seems to conform quite well with the 'law'.

> **TIP:** Before attempting a CPU upgrade it is well worth giving careful attention to the cost effectiveness of the upgrade – in many cases there may be other ways of improving its performance for less outlay. In particular, if you are operating on a limited budget it may be worth considering a RAM or hard disk upgrade *before* attempting to upgrade the CPU. In both cases, significant improvements in performance can usually be achieved at moderate cost.

4.6 Troubleshooting the motherboard

Most motherboard problems are related to cabling and connections. Ensure all cables are connected firmly. Ribbon cables and power cables can often come loose. Ensure all 'plugin' items such as the CPU, RAM modules and adapters such as video cards, modems, etc. are inserted correctly. Contacts can become oxidized or dirty: as a quick fix, remove

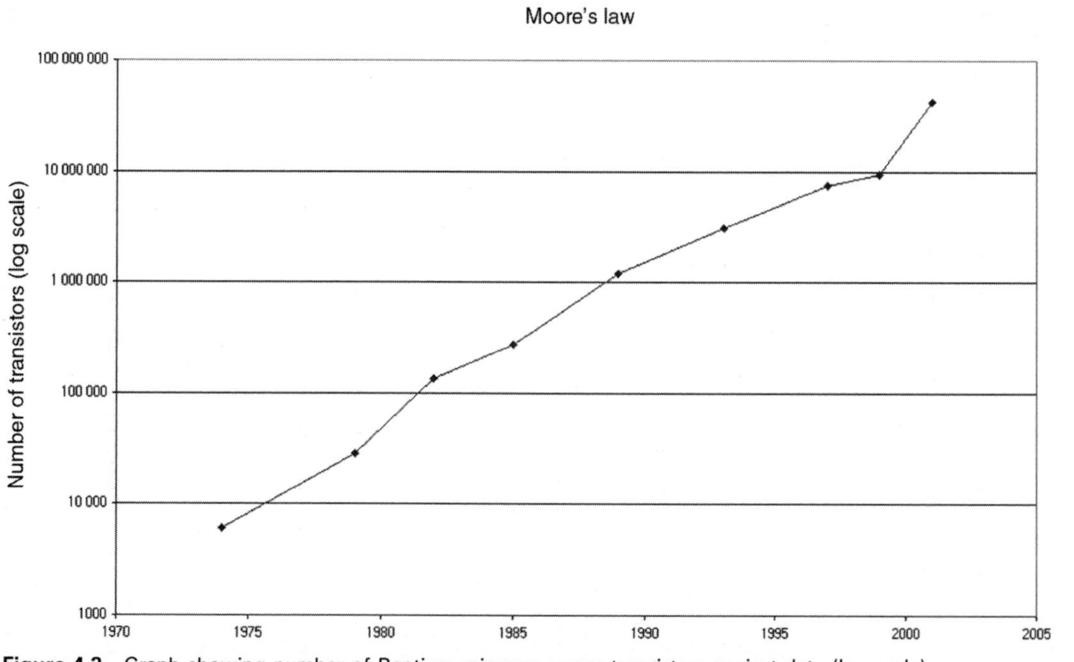

Figure 4.3 Graph showing number of Pentium microprocessor transistors against date (log scale)

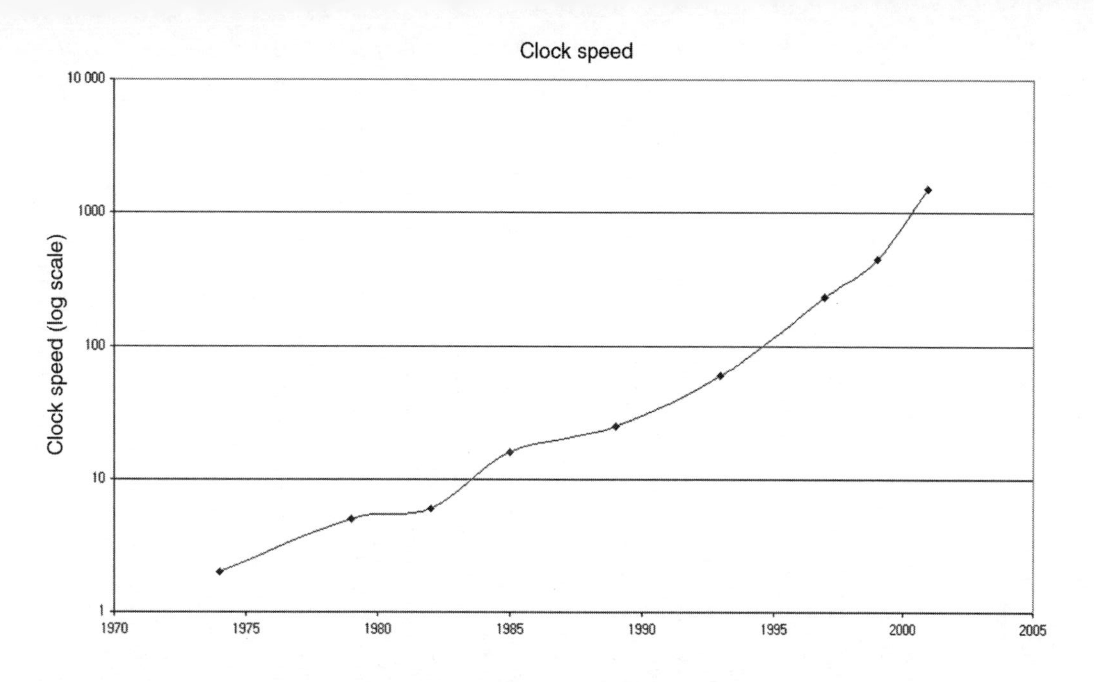

Figure 4.4 Graph showing clock speeds of Pentium microprocessors against date (log scale)

and reinsert the item several times to wear away the oxide. This is not a long-term solution, the parts should be removed and cleaned with a good quality contact cleaner.

Other problems are often related to specific hardware but check the items below first.

General (also refer to Appendix F):

- Remove all add-on cards except the graphics adapter and start the machine. If that fails to give a running machine, check connections, settings, CPU and RAM compatibility, etc. in the motherboard technical specification. Keep the PC speaker connected but not any external speakers.
- Reset the BIOS settings to their default values. On some boards there is a jumper or other way to clear the BIOS settings.
- Is there sufficient power from the power supply?
- Try a different keyboard.
- Check for bent pins on the board. You might be lucky if you try to straighten them. If not, you will need to buy a new board!
- Try disabling the cache in the BIOS. If this makes the machine work, the cache is faulty.

System has no power, no lights come on:

- Check power supply at the wall. Use another mains device to see if that works.
- Check fuse in power lead or try a different lead.
- Remove and firmly reinsert power lead to PC to check if it was loose.
- Using a voltmeter, check that 5 V and 12 V are being supplied from the power supply. In some power supplies there is a fuse on the internal board that may have blown. Great care is needed when working with the inside of the power supply as there is the risk of electrocution from the mains voltage. If the power supply fuse is blown, you must determine why before reuse. If you are unsure, get a technician to have a look or replace the power supply.

System has power, power indicator lights are on, hard drive is spinning:

- Expansion card not fitted firmly.
- Defective expansion card.
- Defective floppy drive.

System boots from the floppy drive, does not boot from hard disk drive (you may get *invalid drive specification* from FDISK):

- Check ribbon cable between hard drive and motherboard.
- Check the drive type in the CMOS setup.
- Damaged hard disk or disk controller.
- Format hard disk; if unable to do so, the hard disk may be defective.

System boots from the floppy drive, does not boot from hard disk drive but hard disk can be read (see also Chapter 10, 'Hard Disk Drives'):

- Boot from floppy with rescue disk. At the DOS prompt type SYS C: to make C: bootable. *Warning,* although this might fix the problem and make C: bootable, any existing operating system software such as Windows will probably need to be reinstalled. At least it rules out a hardware failure. Before any reinstall, back up data files.
- It is probably best to go back to FDISK, re-reset the partitions, format the drive and reinstall the operating system and application. Although this takes quite a long time, a clean install is always easier to work with than a 'fixed' system.

> **TIP:** If you need to make several disk partitions and you are having trouble with FDISK (a very odd program!), the easiest way is to establish at least one partition, format the hard drive and install Windows. Then install Partition Magic to complete the partitioning. This is a painless way to proceed and the instructions with the software are far easier to understand than FDISK! See www.powerquest.com/

Cannot boot system after installing second hard drive:

- Master/slave jumpers not set correctly.
- Hard drives not compatible.

Screen shows 'invalid configuration' or 'CMOS failure':

- Incorrect information in the BIOS. Reboot and enter the BIOS setup (usually by pressing DEL or ALT S or similar at bootup time) then reset default settings.
- Check CMOS battery, the small round silver coloured lithium battery clipped into a mount on the motherboard. If the battery is soldered in place, make a note never to buy hardware from that manufacturer ever again!

> **TIP:** If the CMOS battery works sometimes but not others, remember that batteries give a different voltage with changes in temperature. A warm one may give just enough to work when warm but not enough on a cold day. Such a battery needs replacing.

Screen is blank:

- No power to monitor.
- Monitor not connected to computer correctly.
- Graphics card is loose.
- Check graphics card driver. This is not often the cause of a blank screen, usually if the driver is not correct, you should still see 'something' even if only a flashing cursor or Windows will insist you have a standard VGA monitor, in which case you would get a 640 × 480 pixel, 16 colour screen (as used in Safe mode).

Memory problem as reported by several different error messages:

- Remove all RAM modules, clean contacts and replace in a different order.
- Try RAM from a working PC.
- Try the possibly faulty RAM in a working PC.

Screen goes blank periodically:

- Screen saver is enabled. Turn it off. Screen savers were designed to prevent 'screen burn' where a feint screen image shows permanently when it is off. In these days of planned obsolescence and 'upgrade-itis', it is likely you will buy a new monitor before your current one suffers from screen burn. The machine being used to prepare this book is several years old and is left on all day. There is no sign of screen burn.

Floppy drive lights stays on or 'Error reading drive A:'

- Ribbon cable not connected correctly. Ensure pin1 on the floppy drive corresponds with pin1 on floppy cable connector (usually shown by a coloured stripe on the ribbon cable).
- Faulty drive. Apart from trying it in a working PC, it is not normally worth testing a floppy drive as they are such a low cost item.
- Floppy disk not formatted or formatted for a Macintosh. Macs will read and write PC disks, PCs are weak in this area.

5 The PC expansion buses

The availability of a versatile expansion bus system within the PC must surely be one of the major factors in ensuring its continuing success. The bus is the key to expansion. It allows you to painlessly upgrade your system and configure it for almost any conceivable application.

A number of standards are employed in conventional PC expansion bus schemes, ranging from the original ISA (Industry Standard Architecture) to PCI (Peripheral Component Interconnect) and AGP, the Advanced Graphics Port.

5.1 ISA bus

The PC's ISA bus is based upon a number of expansion 'slots' each of which is fitted with a 62-way direct edge connector together with an optional subsidiary 36-way direct edge connector. The first ISA connector (62 way) provides access to the 8-bit data bus and the majority of control bus signals and power rails while the second connector (36 way) gives access to the remaining data bus lines together with some additional control bus signals. Applications which require only an 8-bit data path and a subset of the PC's standard control signals can make use of only the first connector. Applications which require access to a full 16-bit data path (not available in the early original PC and PC-XT machines) must make use of both connectors. The original PC was fitted with only five expansion slots (spaced approximately 25 mm apart).

ISA slots are provided on some modern motherboards to allow backwards compatibility. Owners of new machines are not always pleased to discover that some older and trusted devices will not fit in their new PC. Some manufacturers have taken the risk of supplying boards with no ISA slots; indeed, it is not likely that ISA slots will be found on new boards for very much longer.

Each board has a requirement to have a separate IRQ line, an interrupt request line. These IRQs are the means that an ISA card uses to announce that it requires attention. Troubleshooting early PCs usually involves resolving these IRQ conflicts. Modern PCs use 'plug and play', a system that sets the IRQs, etc. at bootup time.

5.2 EISA bus

'Extended Industry Standard Architecture' (EISA) is an extension of the ISA bus which has been, and still is, widely supported by a large number of manufacturers. Unlike the ISA bus, EISA provides access to a full 32-bit data bus. To make the system compatible with ISA expansion cards, the standard is based on a two-level connector. The lowest level contacts (used by EISA cards) make connection with the extended 32-bit bus while the upper level contacts (used by ISA cards) provide the 8- and 16-bit connections.

5.3 MCA bus

With the advent of PS/2, a more advanced expansion scheme has become available. This expansion standard is known as 'Micro Channel Architecture' (MCA) and it provides access to the 16-bit data bus in the IBM PS/2 models 50 and 60 whereas access to a full 32-bit data bus is available in the model 80 which has an 80386 CPU. It is now rare to find an MCA motherboard in use.

5.4 VESA (or VL) bus

The Video Electronics Standards Association (VESA), a consortium of over 120 companies, produced the VL bus specification as a solution to the bottleneck imposed by the ISA bus. The VL bus (VESA local bus) allows data to be transferred at much higher speeds than those supported by the ISA and EISA bus standards. It is now rare to find a VESA local bus motherboard in use.

5.5 PCI bus

Initially devised by Intel and subsequently supported by the PCI Special Interest Group (PCI-SIG), the Peripheral Component Interconnect bus has become established as arguably the most popular and 'future proof' bus standard available today. It avoids the IRQ conflicts of the ISA bus by using plug and play.

With plug and play, the system configures itself by allowing the PCI BIOS to access configuration registers on each add-in board at bootup time. As these configuration registers tell the system what resources they need (I/O space, memory space, interrupts, etc.), the system can allocate its resources accordingly, making sure that no two devices conflict. The PCI BIOS cannot directly query ISA devices to determine which resources they need. This can sometimes give rise to problems in systems using both ISA and PCI. A PCI board's I/O address and interrupt are not fixed, and can change every time the system boots.

PCI offers flexible bus mastering. This means that any PCI device can take control of the bus at any time, allowing it to shut out the CPU. Devices use bandwidth as available, even all the bandwidth, if no other demands are made for it. Bus mastering works by sending request signals when a device wants control of the bus and the request being granted if data traffic allows it.

Because the PCI bus is not connected *directly* to the CPU (it is separated by an interface formed by a dedicated 'PCI chipset') the bus is sometimes referred to as a 'mezzanine bus'. This technique offers two advantages over the earlier VL bus specification:

- Reduced loading of the bus lines on the CPU (permitting a longer data path and allowing more bus cards to be connected to it).
- Making the bus 'processor independent'.

The original PCI bus was designed for operation at clock speeds of 33 MHz. With a 32-bit data path, the 33 MHz clock rate implies a maximum data transfer rate of around 130 Mbyte/s (about the same as VL bus). Like the VL bus, the PCI bus connector is similar to that used for MCA. To cater for both 32- and 64-bit operation, PCI bus cards may have either 62 or 94 pins. Later PCI implementations had a bus clock rate up to 66 MHz, giving up to 132 MB per second transfer rate over the 32-bit bus.

Concurrent PCI (supported by modern chipsets) allows for more efficient use of the PCI bus and helps prevent conflicts between PCI and CPU bus mastering devices.

> **TIP:** The PCI bus operates under the control of a separate PCI bus controller (a device within the 'PCI chipset'). The PCI bus thus operates independently of the CPU clock. Thus, when you upgrade your CPU you need not have any concerns about whether your existing PCI cards will cope with the higher clock speed!

The PCI-X specification is a high-performance enhancement to the PCI bus specification. It doubles the maximum clock frequency that can be used by PCI devices from 66 MHz to 133 MHz, thus enabling communication at speeds over 1G byte/s.

On 23 July 2002, the PCI Special Interest Group announced two new specifications: PCI-X 2.0 and PCI Express.

The PCI-X 2.0 specification defines two new versions of PCI-X add-in cards: PCI-X 266 and PCI-X 533.

- PCI-X 266 runs at speeds up to 266 mega transfers per second, enabling sustainable PCI bandwidth of more than 2.1G byte/s.
- PCI-X 533 runs at speeds up to 533 mega transfers per second, enabling bandwidth of more than 4.2G byte/s.

The PCI-X 2.0 specification incorporates ECC, error checking and correction, and is fully backwards compatible with previous generations of PCI. New PCI-X 2.0 adapter cards can be inserted into any PCI slot and operate at the maximum speed of that slot.

PCI Express defines a packet-based protocol with an initial bit rate of 2.5G bit/s aimed at graphics, video editing and streaming multimedia, as well as high-speed interconnects such as USB 2.0, InfiniBand and Gigabit Ethernet. It supports hot swapping.

5.6 Troubleshooting the PCI bus

Trouble with PCI devices can be caused by

- Software bugs.
- Software settings.
- Hardware faults.
- Device conflicts.

The standard way to fix software bugs is to obtain the latest card driver from the manufacturer. This can usually be achieved via the manufacturers' website. These websites often contain details of hardware or software conflicts.

TIP: If you need to upgrade a device driver, it is often better to uninstall the old one first. Reinstallation sometimes retains old (and possibly faulty) software components such as .DLL files.

Some boards will not work if they are present in the same machine as other devices. An example of this occurred in the PC of one of the authors. It was fitted with a PCI SCSI card that allowed a SCSI scanner to work very well. All was well until the scanner and its SCSI cable were removed. After this, the IDE CD-ROM drive eject button caused the system to reboot. Removing the now unused PCI SCSI card failed to resolve the problem. It had to be reinserted, the driver removed via the Windows 98 Control Panel and the card removed again before a reboot caused the 'new hardware found' dialogue. Problems with seemingly unrelated devices are not uncommon. This problem was discovered via the Microsoft knowledge base website (support.microsoft.com/).

Only after trying to resolve software/hardware conflicts should a hardware fault be suspected. If a card is suspected of giving trouble, shut down the system, remove the card and install it in a second PC. If the trouble persists in the second PC, the card is probably faulty – repair is not usually economic.

Without specialist equipment, troubleshooting the PCI can be tricky. Test equipment such as the PCI diagnostics card from UltraX Inc. (www.uxd.com/phdpci.shtml) will test the PCI bus when other devices are dead or missing.

5.7 Accelerated graphics port (AGP)

Accelerated graphics port (AGP) is a bus specification for graphics adapters. The main use is to enable 3-D graphics to display quickly on PCs.

Graphical techniques such as texture mapping, z-buffering and alpha blending are all very calculation intensive tasks. If these tasks were performed by the CPU and the resulting data moved over the PCI bus to/from RAM, the PC would provide a speed of operation too slow for 3-D games, etc. The AGP is a separate bus from the PCI bus so can communicate with the CPU directly at its own speed.

The PCI bus currently runs at 33 MHz which allows it to achieve a maximum data transfer rate of 132 M byte/s whereas AGP can achieve a peak transfer rate of 528 M byte/s.

Older software cannot make use of AGP ports as the software must be written to 'know' about AGP. All current machines with Direct X drivers will support AGP graphics cards.

5.8 Troubleshooting the AGP

Most trouble associated with the AGP has nothing to do with the AGP itself, it is more likely to be with the graphics card or the monitor, especially with software/driver settings:

- The computer doesn't boot after new video card is installed. Another video graphics card may be in conflict with the new card, so remove the other one.
- The screen image is off-centre, the colours are wrong or there is no picture at all. This is more likely to be caused by monitor settings or the monitor cable. Check the settings and for bent/broken pins in the video lead. If your monitor uses BNC connectors, ensure that each of the red, green and blue lines are connected and that the vertical sync wire (black) is connected to the correct connector.
- After the Windows startup screen appears, the screen image is garbled or unusable. The driver may be using settings your monitor doesn't

support. This will occur if Windows failed to correctly identify the monitor type. Reboot in safe mode by pressing F8 after the initial BIOS screen. As an experiment, set the display type to 'VGA', 640 × 480 pixels, 256 colours, a setting that will work with virtually any display. If that setting works there is unlikely to be a hardware fault. The next step is to set the display resolution to that claimed by the manufacturers. If that fails, go back to safe mode and reset the screen vertical refresh rate a littler slower. 75 Hz should give a steady image, any faster and the monitor may not be able to 'keep up'.

- After a Windows 95 game starts, the monitor doesn't display properly. If your game uses a low-resolution (640 × 480 pixels or below), full-screen display mode, your monitor may not support the refresh rate that the MGA driver is using.
- After you restart the computer, Windows reports that the graphics card is not configured correctly. This may be caused by a previously installed display driver. Delete all drivers and reinstall.

6 Semiconductor memory

This chapter will help you understand the use of RAM and ROM within your PC. It also explains how you can locate and replace a faulty memory chip.

6.1 Memory basics

Semiconductor memory devices tend to fall into two main categories; 'read/write' and 'read only'. Read/write memory is simply memory which can be read from and written to. In other words, the contents of the memory can be modified at will. Read-only memory, on the other hand, can only be read from; an attempt to write data to such a memory will have no effect on its contents.

6.1.1 Memory organization

Each location in semiconductor ROM and RAM has its own unique address. At each address a byte (comprising 8 bits) is stored. Each ROM, RAM (or bank of RAM devices) accounts for a particular block of memory, its size depending upon the capacity of the ROM or RAM in question.

6.1.2 BIOS ROM

The BIOS ROM that contains the low-level code required to control the system's hardware is programmed during chip manufacture. The programming data is supplied to the semiconductor manufacturer by the BIOS originator. Older BIOS ROMs were cost effective for large-scale production; however, programming of the ROM is irreversible – once programmed, devices cannot be erased in preparation for fresh programming. Hence, the only way of upgrading the older BIOS ROMs is to remove and discard the existing chips and replace them with new ones. More modern Flash BIOS can be overwritten with new or updated BIOS software. This is an area that has been attacked by those sick people who write viruses. This is one area in which older technology actually shows some benefits!

6.2 Upgrading your BIOS ROM

At some point, you may find it necessary to upgrade the BIOS ROM on a machine. There are various reasons for doing this but most centre on the need to make your software recognize significant hardware upgrades (e.g. to make use of an IDE hard drive or when replacing the 360 K or 720 K floppy drives on an older machine with newer 1.44 M or 2.88 M drives).

With a DIL packaged BIOS ROM (either socketed or soldered in) the following stages are required:

1. As far as possible, make sure that the new BIOS is compatible with your system (it might be wise to ask the supplier if he/she will offer a refund if you have any problems).
2. Note down your existing CMOS configuration (using your 'setup' program).
3. Power down your system.
4. Locate the BIOS ROM chips (note their position and orientation).
5. Remove and replace the BIOS ROM chips.
6. Reassemble the system and run the 'setup' program, making any changes necessary.

TIP: In a two-chip BIOS ROM set, the chips are usually marked 'Low' and 'High' (or 'Odd' and 'Even') in order to distinguish them from one another. Always make sure that you replace them in the correct sockets (i.e. locate the new 'Low' chip in the socket vacated by the old 'L' chip).

Some of the latest generation of motherboards have their BIOS stored in electrically erasable read-only memory (EEPROM). This memory can be easily reprogrammed *without* having to remove the BIOS chip(s) from the motherboard.

6.2.1 Flash memory

Flash memory makes upgrading your BIOS easy. A new version can be installed from a disk supplied by the manufacturer. Alternatively, BIOS upgrades can be distributed through the Internet (simply download them from the OEM's site and then run the executable file). The Intel disk-based flash upgrade utility (FMUP.EXE) has three options:

- The flash BIOS can be upgraded from the disk.
- The current BIOS code can be copied from flash memory onto the disk (for backup purposes).

- The data in the flash memory can be compared with that on the disk to determine whether or not the current version is installed.

Before operation, the upgrade utility must first check that the system's hardware (the 'target system') is fully compatible with the BIOS upgrade. This helps to avoid the danger of installing a BIOS upgrade intended for a different hardware configuration!

6.3 Random access memory (RAM)

RAM stands for random access memory and must rank as the silliest name in computing. Many different storage devices can be accessed randomly. A better name would be volatile memory because the data stored in RAM is lost when the power is lost.

RAM is used to make a temporary if high-speed storage area for program instructions and data. Computers work by fetching then executing instructions from RAM so the speed of RAM is a very important factor in the performance of the whole machine. RAM is nowhere near as fast as modern CPUs so extensive use of cache memory is made. As far as the CPU is concerned, program instructions and data are supplied very quickly from the cache but the RAM must be able to supply blocks of data to the cache at high speed.

Modern RAM is supplied in various sizes of miniature circuit board or module; older PCs had dedicated RAM chips on the motherboard. These miniature circuit boards are called SIMM for single inline memory module, DIMM for dual inline memory module and variations of these terms.

Table 6.1 RAM board types

Name	Typical usage	Voltage	Speed
30-pin SIMM	286 and 386 PCs	5 V	60 ns to 80 ns
72-pin SIMM	486 and Pentium PCs	3.3 V and 5 V	60 ns to 70 ns
168-pin SIMM	Most modern PCs	SDRAM 3.3 V	66 MHz, 100 MHz and 133 MHz
168-pin DIMM	Most modern PCs	EDO/FPM 3.3 V and 5 V	66 MHz, 100 MHz and 133 MHz
144-pin SODIMM	Laptops	3.3 V	
184-pin RIMM	Latest Intel RamBus motherboards	2.5 V	600 MHz, 700 MHz, 800 MHz, 1 GHz

Historical note: Speed quoted in ns means nanoseconds or 10^{-9} seconds. In the days when the microprocessor was connected more or less directly with RAM, the cycle time of the fetch–execute sequence between the microprocessor and its RAM was important. If the microprocessor had a cycle time of 60 ns, a RAM of 80 ns would be too slow. Modern PCs use cache memory and other techniques to deal with any mismatch in the RAM/microprocessor speeds.

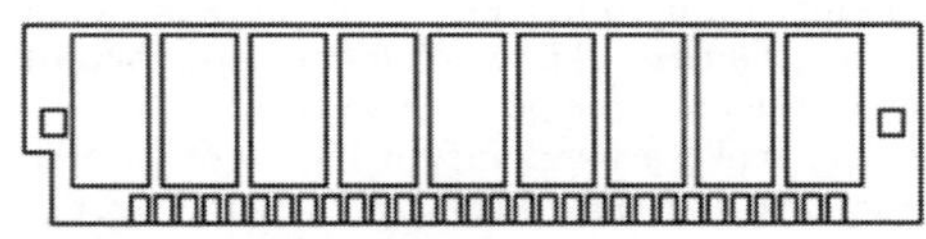

Figure 6.1 30-pin SIMM

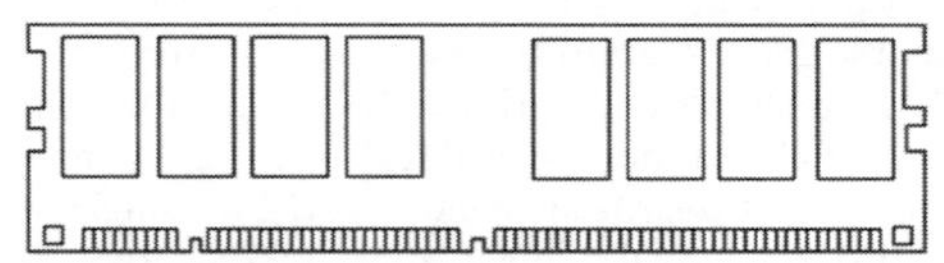

Figure 6.2 168-pin DIMM

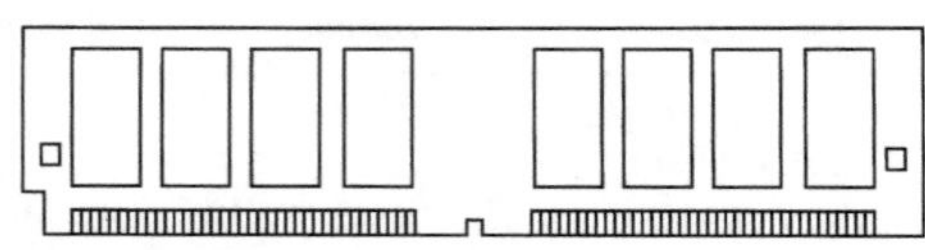

Figure 6.3 72-pin DIMM

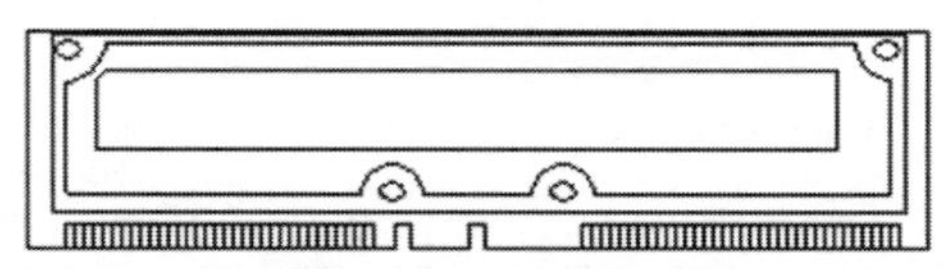

Figure 6.4 184-pin RIMM for RamBus

Apart from different packaging such as SIMMs or DIMMS, there are different RAM types, each generally faster than the last.

DRAM (or dynamic RAM) is the oldest type. It stores each bit in a storage cell as an electrical charge that must be refreshed every few milliseconds to retain the information.

Static RAM, unlike DRAM, does not require refreshing but is expensive.

EDO RAM (or extended data out RAM) is faster than DRAM and extends the time that output data is valid. EDO RAM has now been replaced by SDRAM.

FPM (or fast page mode RAM) is a kind of DRAM that allows faster access to data in the same row or page. Superseded by SDRAM.

SDRAM or synchronous DRAM has replaced DRAM, FPM and EDO and is currently the most common type in modern machines. SDRAM is able to synchronize memory access with the CPU clock. This allows one whole block of data to be sent to the CPU while another is being prepared for access. Typical speeds are 66 MHz, 75 MHz, 100 MHz and 133 MHz.

The 184-pin RIMM is used in the latest Intel PCs. It has a high operating frequency with speeds of 600, 700 and 800 MHz.

Table 6.2

RAM type	Alternative name	Speed	DDR speed	Voltage	Bandwidth: GB/second
PC100		100 MHz		3.3 V	0.80
PC133		133 MHz		3.3 V	1.05
PC1600	DDR200	100 MHz	200 MHz	2.5 V	1.60
PC2100	DDR266	133 MHz	266 MHz	2.5 V	2.10
PC2700	DDR333	166 MHz	333 MHz	2.5 V	2.70
PC3200	DDR400	200 MHz	400 MHz	2.5 V	3.20
PC4200	DDR533	266 MHz	533 MHz	2.5 V	4.20
RDRAM PC800		400 MHz			1.60
RDRAM PC1066		533 MHz			2.10
RDRAM PC1200		600 MHz			2.40

6.4 RAM troubleshooting

- Check speed and type against the motherboard technical specification. Some people claim that mixing RAM types is OK but not all systems will accept that.

- Insert the old RAM without the new RAM to see if other problems exist. The new RAM may be OK but of the wrong type.
- Fitting RAM of the wrong speed may give intermittent faults.
- A loose RAM module will also give intermittent faults.
- Make sure the RAM modules are latched into place on the motherboard.
- Remove the RAM, use a proprietary contact cleaner with cotton buds then refit the RAM modules. If you have no access to a good cleaner, simply remove and replace the modules a number of times. This will wear through any oxide or detritus but is only a temporary fix, the contacts should be cleaned.
- Check your PC's CMOS setup, the setting made need to be changed. Sometimes, if you save the current settings then exit the CMOS setup will force it to recalibrate and find all of the installed RAM.
- If there are two or more RAM modules, fit them in a different order to see if you get the identical fault. If not, something else is wrong; if swapping the order changes the problem, probably only one of the modules is faulty. Some BIOSs will give a message that some RAM is bad. Be aware that some boards require RAM modules in pairs.
- The memory may be defective. Try the modules in another compatible PC to see if they work.
- If the PC works well for a while then crashes, suspect a cooling problem. Try turning it off for an hour then restarting. If it works then crashes after warming up, have a careful look at cooling. You may find that when you fitted the RAM that you moved some ribbon cables. These may have blocked the airflow over the RAM or the CPU.
- Check that when you fitted the RAM that nothing else was disturbed, especially ribbon cables that may have been pulled from their seatings.
- If the PC gives an unusual number of beeps during bootup, refer to Appendix F.

If you see that the contacts on the motherboard are bent or faulty, you may be lucky and have spare sockets you can use. Repairs are usually not economical so a new motherboard will fix the problem!

A technique suggested by Microsoft to help diagnose RAM troubles can be found at
support.microsoft.com/default.aspx?scid=kb;en-us;142546.

A 'memory leak' is a software bug and nothing to do with hardware. Sometimes you will see an error message from Windows saying a 'memory leak' has occurred or that 'there is not sufficient memory available'. Neither of these messages are caused by hardware faults, they are usually caused by careless programming. If you reboot the machine and it appears to work, this more or less proves the point that the problem lies with software. See the section on software trouble-shooting.

6.5 CMOS RAM

This is not a type of RAM but is the memory that stores machine specific data such as the time of the real-time clock, kinds of hard drive fitted, etc.

It is powered by a small lithium battery, usually on the motherboard. Unless it is soldered in place, replacement is simple. When the CMOS battery fails or when power is inadvertently removed from the real-time clock chip, all data will become invalid and you will have to use your setup program to restore the settings for your system. It is a good idea to note down the CMOS settings, especially in older machines fitted with newer components in case you have to manually restore them after a CMOS setting failure.

If you have a power-on password set, this is held in the CMOS memory. Removal of the battery for a short while will force a reset of the settings to the manufacturer's defaults so overcoming the password. For this reason you should not rely on such a password!

6.6 Memory diagnostics

6.6.1 ROM diagnostics

The PC's BIOS ROM incorporates some basic diagnostic software which checks the BIOS ROM and RAM during the initialization process. The ROM diagnostic is based upon a known 'checksum' for the device. Each byte of ROM is successively read and a checksum is generated. This checksum is then compared with a stored checksum or is adjusted by padding the ROM with bytes which make the checksum amount to zero (neglecting overflow). If any difference is detected an error message is produced and the bootstrap routine is aborted.

6.6.2 RAM diagnostics

In the case of RAM diagnostics the technique is quite different and usually involves writing and reading each byte of RAM in turn. Various algorithms have been developed which make this process more reliable (e.g. 'walking ones'). Where a particular bit is 'stuck' (i.e. refuses to be changed), the bootstrap routine is aborted and an error code is displayed. This error code will normally allow you to identify the particular device that has failed.

More complex RAM diagnostics involve continuously writing and reading complex bit patterns. These tests are more comprehensive than simple read/write checks. RAM diagnostics can also be carried out on a non-destructive basis. In such cases, the byte read from RAM is replaced immediately after each byte has been tested. It is thus possible to perform a diagnostic check some time after the system has been initialized and without destroying any programs and data which may be resident in memory at the time.

6.6.3 Software for checking RAM

Much software is available, either shareware, freeware or full commercial software. A recent search for a 'memory diagnostic' on www.tucows.com gave 115 titles.

6.6.4 Parity checking

The integrity of stored data integrity can be checked by adding an extra 'parity bit'. This bit is either set or reset according to whether the number of 1s present within the byte are even or odd (i.e. 'even parity' and 'odd parity').

Parity bits are automatically written to memory during a memory write cycle and read from memory during a memory read cycle. A non-maskable interrupt (NMI) is generated if a parity error is detected and thus users are notified if RAM faults develop during normal system operation.

TIP: Parity errors can very occasionally occur due to the spontaneous passage of stray radioactive particles through a RAM chip. If this phenomenon does occur, and your system reports a 'parity error' and then shuts down, it will usually reboot. This type of error is often referred to as a 'soft error' and it will not normally recur. Repeated or permanent parity errors, on the other hand, usually indicate a failed (or failing) RAM chip. These 'hard errors' usually mean that you must replace a chip or module to restore normal operation.

7 Printers and the printer interface

The PC's parallel ports (LPT1 and LPT2) provide a very simple and effective interface which can be used to link your PC to a wide range of printers and other devices such as external tape and disk drives. This chapter explains the principles of parallel I/O and describes the Centronics interface standard before discussing basic fault finding and troubleshooting procedures which can be applied to the parallel interface.

7.1 Parallel I/O

Parallel I/O is used to transfer bytes of data at a time between a microcomputer and a peripheral device (such as a printer). Several control signals are present in order to achieve 'handshaking', the aptly named process which controls the exchange of data between the computer system and the printer.

A basic handshake sequence is as follows:

1. The PC indicates that it is ready to output data to the printer by asserting the STROBE line.
2. The PC then waits for the printer to respond by asserting the ACK (acknowledge) line.
3. When ACK is received by the PC, it places the outgoing data on the eight data lines.
4. The cycle is then repeated until the printer's internal buffer is full of data.

The buffer may have to be filled several times during the printing of a large document. Each time, the port will output data at a fast rate but the printer will take an appreciable amount of time to print each character and thus will operate at a very much slower rate. Clearly, your PC will be 'tied up' for less time if you have a larger printer buffer!

7.2 ECP/EPP (Centronics) printer port

The Centronics interface has become established as the most commonly used interface standard for the transfer of data between a PC and a printer. The standard employs parallel data transmission (a byte is transferred at a time).

The standard is based on a 36-way Amphenol connector as in Figure 7.1. The interface is generally suitable for transfer of data at distances of up to 4 m, or so.

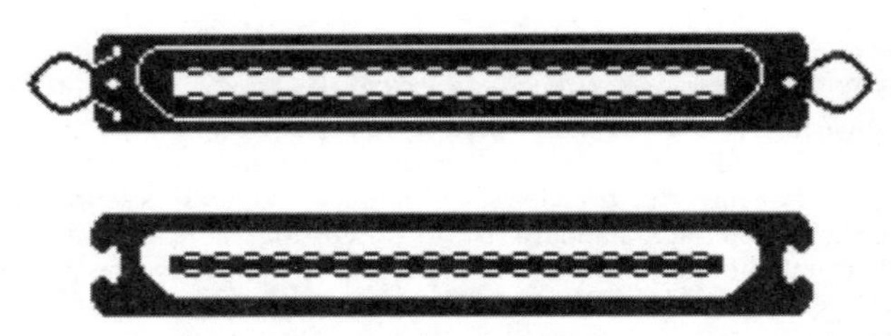

Figure 7.1 Centronics Amphenol 36-way male and female connectors

It is also worth noting that data transfer is essentially in one direction only (from the microcomputer to the printer). Indeed, some early PCs have printer ports which can only be configured in one direction (i.e. output only).

7.3 Printer types and emulations

When a computer sends data to a printing device, the data may take one of several forms. This affects the way that programmers must think about how they are to format their data and how to make the best use of the printer's differing functions; it also affects the speed of response as seen by the user. More complex print formatting requires more sophisticated communication with the printer. When you install software on a machine, one task is to load a *printer driver*. This piece of software has the task of taking the data from an application including any markup that defines formatting, etc. and translating it to the form the printer will accept. Many modern printers will accept all the forms shown below.

Older printers would send plain ASCII codes to the printer with escape sequences. This method uses special codes to tell the printer when to underline, print double high, etc. These codes use ASCII character 27 (1B in hex), the *escape* character. For instance, if the word 'CAT' is sent to the printer, the ASCII codes would be 67, 65, 84 (or 43, 41, 55 in hex), the codes for C, A and T. If you wished to have this underlined you would first send the escape sequence 27, 45, 1 that is 'ESC', '-', '1'.

Other examples of escape sequences:

ESC, 'W', 1 turn on double wide printing, ESC, 'W', 0 to turn it off
ESC, '4', 1 turn on italic printing, ESC, '5' to release
ESC, '@' resets the printer

More modern printers use a *page description language*. This is when code is used to *describe* the page layout and contents then send that description to the printer. The printer usually contains its own computer which interprets the description and forms an image to be printed. This computer is sometimes more powerful than the PC used for word-processing, etc.!

The page description could be either in the Adobe *PostScript* language or the Hewlett Packard language called *PCL*. However the page is described, it must be *rasterized*, i.e. turned into a set of dots; this is the purpose of the computer in the printer. It is possible to do this in the main PC but this slows down the PC as far as the user is concerned and generates huge amounts of data to be downloaded to the printer, further slowing down the process.

Windows: Some printers are designed for use by PCs running Windows only, they will not work on other operating systems. This is to make the printer cheaper. The PC will rasterize (meaning to convert the text and graphics to rows of dots ready for printing) the data for printing, eliminating the need for a powerful CPU and RAM in the printer. When printing graphics using Windows, you may see the PC appear to slow down. It has not slowed down, it is busy!

The advantage gained by using a page description language is much better control of the printed image. For instance, the original software does not need to know how to place each dot needed to form a line, only where the line starts and ends, the printer works out the placement of each dot in the line, i.e. it generates the whole line from just the end point data. This also means that the printer can hold the shape of each letter (or *glyph*) in each size (called *fonts*), so the image of each letter does not need to be sent to the printer. If unusual letters are required, they can be sent to the printer hence allowing almost unlimited printing of characters and graphics images.

PCL is based on or evolved from an advanced series of escape sequences but PostScript is a stack-based computer language that may be written 'by hand' or interpreted by other software. You could if you wished write programs in PostScript. Software is available (such as Ghost Script) that will form an on-screen image of the document just by using the PostScript information. A feature of PostScript is that it is not dependent on the printer resolution. This means that if you send a PostScript file to a low resolution printer it will be rendered according to that printer. If you send the same file to a high resolution machine you get the same image in the same proportion but looking much sharper. This is the method of choice for many DTP operations. The image is composed on a PC and checked on a local laser printer but the file is sent to a professional typesetter who would use a printer with many times the resolution.

The use of a page description language allows much better control of the printed image and is most suitable for everyday office use, but don't forget that not all printers are attached to PCs or bigger machines. The printers in point of sale machines that produce till receipts still need to be controlled as do numerous other small printing devices and many of these are still dot matrix printers. Many businesses choose to keep their dot matrix printers for printing invoices instead of using laser printers; they are simple, robust and very fast for small print jobs because they do not have a long startup time.

7.4 Troubleshooting the printer

Printer cables and connectors often prove to be troublesome (particularly when they are regularly connected and disconnected) and it is always worth checking the cable first.

Refer to the printer handbook to determine how to print a self-test page. This is often achieved by holding down a key while turning the machine on or choosing a menu item on a small screen.

It will usually be a fairly easy matter to decide which part of the interface (printer, cable or the PC's parallel port) is at fault. Where text is printed but characters appear to be translated resulting in garbled output, one or more of the data line signals may be missing. In this case it is worth borrowing another printer cable.

Where the handshake signals are missing, you will usually be warned by an on-screen error message (such as 'printer not responding' or 'printer off-line'). Many software packages are supplied with printer configuration files or 'printer driver' files which ensure that the control codes generated by the software match those required by the printer.

7.5 Printing from Windows

If you use the sequence Start → Settings → Printers then right click on a printer icon, Windows will allow you to print a test page. In the later versions of Windows, there is a printer troubleshooter that is sometimes helpful.

TIP: One very annoying 'feature' of the Windows printing system is that if you pause a printer then forget you have done it, it appears the thing has gone wrong. No message shows what you have done unless you remember to use Start → Settings → Printers, then right click on a printer icon, then unpause it. Always check the printer icon before suspecting anything else.

TIP: If you suspect that the Windows printer driver software is at fault, always remove the driver before installing a new one. Some installation routines keep the possibly fault settings already on the PC.

7.6 General printer troubleshooting

- Ensure the printer has power and is on-line.
- Ensure any printer sharer switch is set correctly.
- Turn printer off and on again and try again.
- Ensure the ink cartridge seal is removed.
- Try the printer's own self-test.
- In Windows, use the sequence Start → Settings → Printers and check if the print job is waiting.
- Check if the correct printer driver is installed.
- Replace the cable with a known good cable.
- Check the total length of the cabling from the PC to the printer. Distances of up to 4 metres can be supported but 3 metres is better. The parallel interface will not work reliably with longer cables.
- Try connecting a different printer to this PC.
- Try connecting the printer to a PC known to work (but with the correct driver).

The printed colours do not match screen colours:

- Check colour profiles set in your printer driver. This topic is too complex to cover in a book of this kind and is a cause of much trouble when precise colours are required. One problem is that the screen is bright making it look like a backlit picture, printed images are much duller. See www.color.org/profile.html

Poor print quality:

- Match the paper to your printer. Laser printers will take most paper types but inkjets will not.
- Paper has a high moisture content or a rough surface.

Printer prints blank pages:

- Empty or defective toner or ink cartridge.
- Sealing tape or tab left in toner cartridge.

Printer prints black pages or thin, dark, vertical black lines:

- Improperly installed or defective toner or ink cartridge.

Image is skewed:

- Paper is loaded incorrectly.

Printer prints out of line on the paper (poor registration):

- Paper is too light or too heavy.
- Paper is loaded incorrectly, or guides are misadjusted.
- Leading edge of paper is curled.
- Paper tray is overloaded.

Light print or faded print:

- Paper is out of specification.
- Low or empty toner or ink cartridge.
- Print density improperly set.

Printer prints dots on what should be blank areas:

- Print density improperly set.
- Defective toner or ink cartridge.
- Media is out of specification.
- Wet paper.
- Inside of printer is dirty.

Back of page is dirty:

- Toner has leaked from cartridge.
- Inside of printer is dirty.

Thin, vertical white lines/stripes:

- Toner cartridge is nearly empty or defective.
- Printer is dirty.

Portion of page is blank:

- Page layout is too complex.
- Not enough memory.
- Printing on legal size paper when software specifies letter size.

8 The serial communication ports

The PC's serial communication ports (COM1, COM2, etc.) were once the means of linking a PC with the rest of the world. Modern machines use much faster connections such as USB but there are many *legacy* (meaning old) devices that use the serial port to connect to the PC. Usually, these devices do not need great speed, an example would be a GPS (global positioning system).

The serial port is made to an old computing 'standard'. As has been said many times, the nice thing about computer standards is that there are so many to choose from! The serial port standard is known as RS-232C, RS means 'recommended standard'. Older RS-232 devices were seldom 'standard' and are best left in a museum. RS-232 uses a 25-way D-type connector, but most serial ports on modern machines use a 9-way D-type connector and now achieve a remarkable degree of standardization. But not always. In the possession of one of the authors of this book is a list of over 70 'standard' cable wiring diagrams, used to connect diverse devices. In difficult cases, it may still be necessary to use a *breakout box*, a simple device that allows experimental connections of various RS-232 pins until a combination is found that works.

As can be seen from Figures 8.5 to 8.10, several different cables exist even off the shelf to suit the idiosyncrasies of serial devices. Apart from speed, this is one good reason to use USB devices.

8.1 The RS-232 standard

The RS-232 standard was first defined by the Electronic Industries Association (EIA) in 1962 as a recommended standard (RS) for modem interfacing. The latest revision of the RS-232 standard (RS-232D, January 1987) brings it in-line with international standards CCITT V24, V28 and ISO IS2110. The RS-232D standard includes facilities for 'loop-back' testing which were not defined under the previous RS-232C standard.

8.1.1 Terminology

The standard relates essentially to two types of equipment: data terminal equipment (DTE) and data circuit terminating equipment (DCE). data terminal equipment (i.e. a PC) is capable of sending and/or receiving data via the COM1 or COM2 serial interface. It is thus said to 'terminate' the serial link. Data circuit terminating equipment (formerly known as data communications equipment), on the other hand, facilitates data communications. A typical example is that of a 'modem' (modulator–demodulator) which forms an essential link in the serial path between a PC and a telephone line.

TIP: You can normally distinguish a DTE device from a DCE device by examining the type of connector fitted. A DTE device is normally fitted with a male connector while a DCE device is invariably fitted with a female connector.

TIP: There is a subtle difference between the 'bit rate' as perceived by the computer and the 'baud rate' (i.e. the signalling rate in the transmission medium). The reason is simply that additional start, stop and parity bits must accompany the data so that it can be recovered from the asynchronous data stream. For example, a typical PC serial configuration might use a total of 11 bits to convey each 7-bit ASCII character. In this case, a line baud rate of 600 baud implies a useful data transfer rate of a mere 382 bits per second.

8.1.2 The RS-232 connector

A PC serial interface is usually implemented using a standard 25-way D connector (see Figure 8.1). The PC (the DTE) is fitted with a male connector and the peripheral device (the DCE) normally uses a female connector. When you need to link two PCs together, they must *both* adopt the role of DTE while thinking that the other is a DCE. This little bit of trickery is enabled by means of a 'null modem'. The null modem works by swapping over the TXD and RXD, CTS and RTS, DTR and DSR signals.

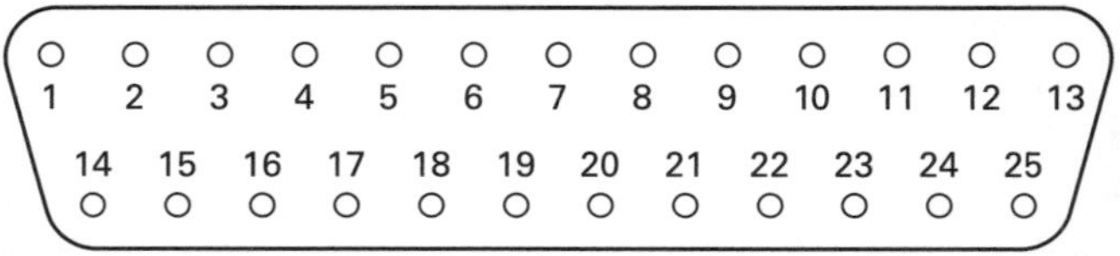

Figure 8.1 25-way D-type connector, PIN side

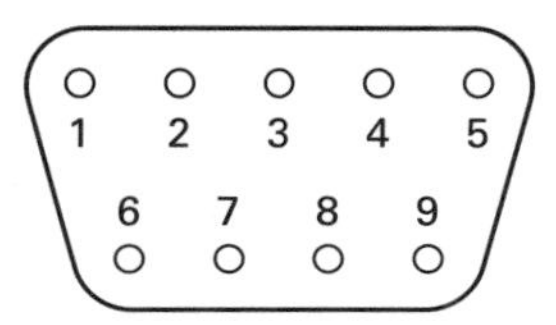

Figure 8.2 9-way D-type connector, PIN side

TIP: There are various types of RS-232 data cable. Some may have as few as four connections, many have nine or 15, and some have all 25. When you purchase a cable it is worth checking how many connections are present within the cable. A cheaper 9-way cable will *usually* work provided your software does not make use of the 'ring indicator' facility.

8.2 Troubleshooting the serial ports

- Troubleshooting the communication ports usually involves the following basic steps. Remember, the biggest problem with legacy serial ports are differences in cable connections and software settings.
- Check the physical connection between the PC (the 'DTE') and the DCE (e.g. the modem). Where both devices are PCs (i.e. both configured as DTE) a patch box or null modem cable should be used for correct operation.
- Check that the same data word format and baud rate has been selected at each end of the serial link (note that this is most important and it often explains why an RS-232 link fails to operate even though the hardware and cables have been checked). This is done in whatever software you are using on your PC.
- Activate the link and investigate the logical state of the data (TXD and RXD) and handshaking (RTS, CTS, etc.) signal lines using a line monitor, breakout box, or interface tester. Lines may be looped back to test each end of the link.
- If in any doubt, refer to the equipment manufacturer's data in order to ascertain whether any special connections are required and to ensure that the interfaces are truly compatible. Note that some manufacturers have implemented quasi-RS-232 interfaces which make use of TTL signals. These are *not* electrically compatible with the normal RS-232 system even though they may obey the same communication protocols. They will also *not* interface directly with a standard PC COM port!
- The communications software should be initially configured for the 'least complex' protocol (e.g. basic 7-bit ASCII character transfer at

V.24/RS-232 connections

Source	Designation	Pin
DTE	Secondary transmit data	14
DCE	Transmit clock (DCE source)	15
DCE	Secondary received data	16
DCE	Receive clock	17
DTE	Local loopback LL	18
DTE	Secondary request to send	19
DTE	Data terminal ready DTR	20
DTE	Remote loopback RL	21
DCE	Ring indicator RI	22
DTE/DCE	Baud rate select	23
DTE	Transmit clock (DTE source)	24
DCE	Test mode	25

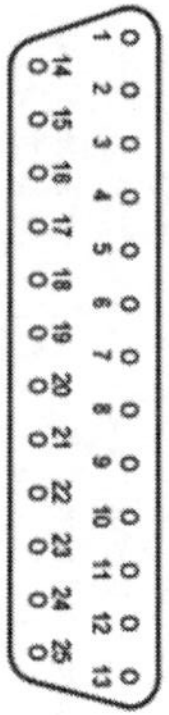

Pin	Designation	Source
1	Protective Ground	Common
2	Transmit data TX	DTE
3	Receive data RX	DCE
4	Request to send RTS	DTE
5	Clear to send CTS	DCE
6	Data set ready DSR	DCE
7	Signal ground	Common
8	Carrier detect	DCE
9	Reserved +V	n/a
10	Reserved -V	n/a
11	Unassigned	n/a
12	Secondary carrier detect	DCE
13	Secondary clear to send	DCE

Figure 8.3 25-way RS-232 connections

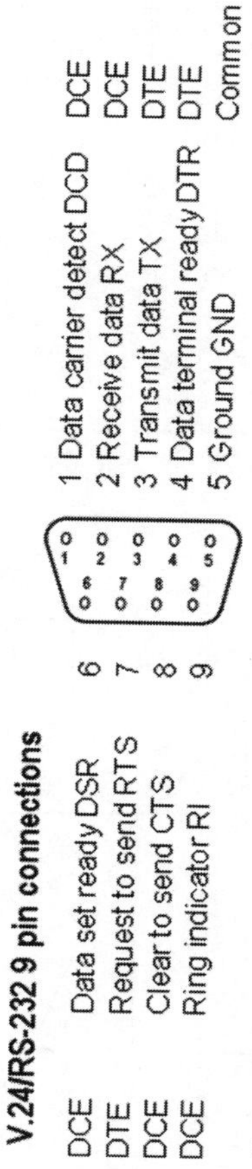
V.24/RS-232 9 pin connections
DCE Data set ready DSR 6
DTE Request to send RTS 7
DCE Clear to send CTS 8
DCE Ring indicator RI 9
1 Data carrier detect DCD DCE
2 Receive data RX DCE
3 Transmit data TX DTE
4 Data terminal ready DTR DTE
5 Ground GND Common
1 2 3 4 5
6 7 8 9

Figure 8.4 9-way RS-232 connections

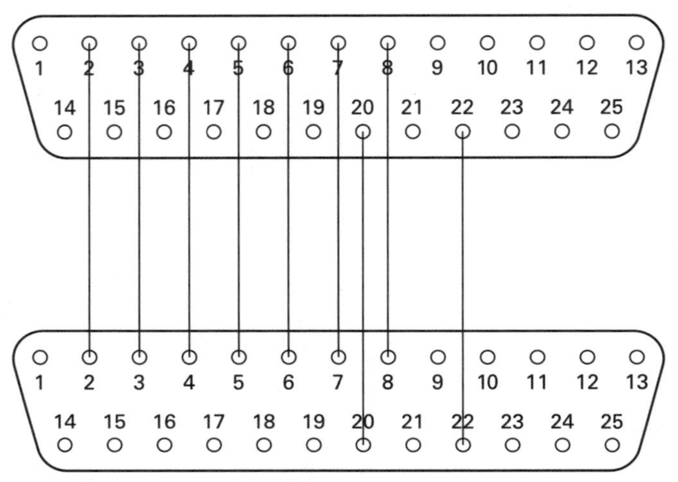

Figure 8.5 25-pin DTE to 25-pin DCE cable connection

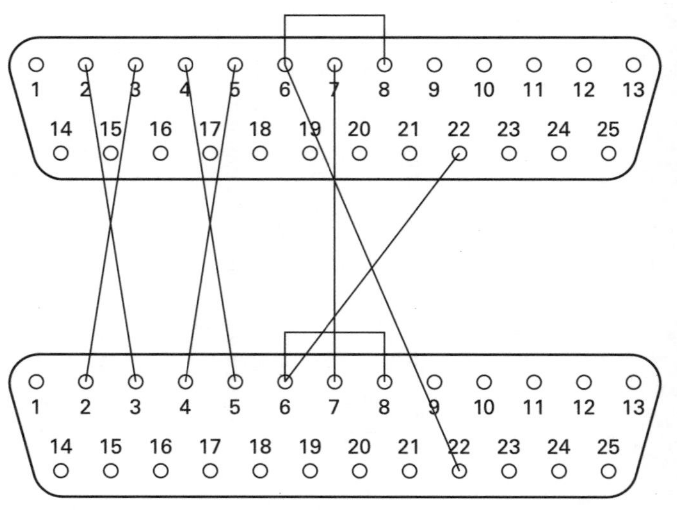

Figure 8.6 25-pin DTE to 25-pin DTE null modem cable connection

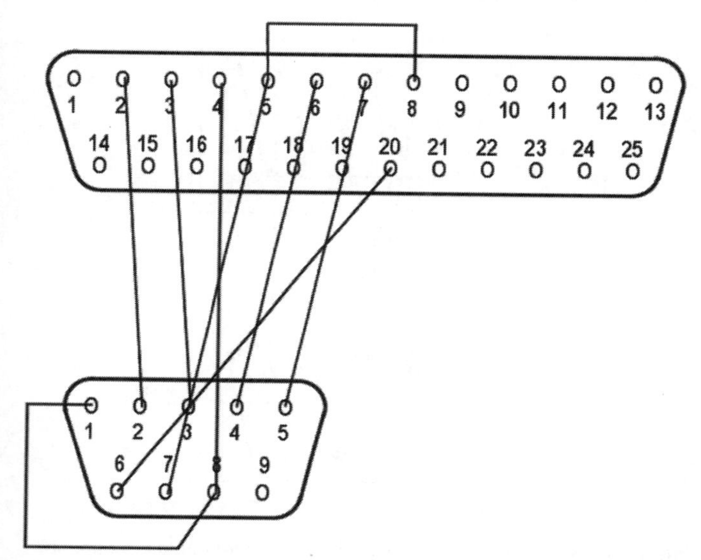

Figure 8.7 25-pin DTE to 9-pin DTE null modem cable connection

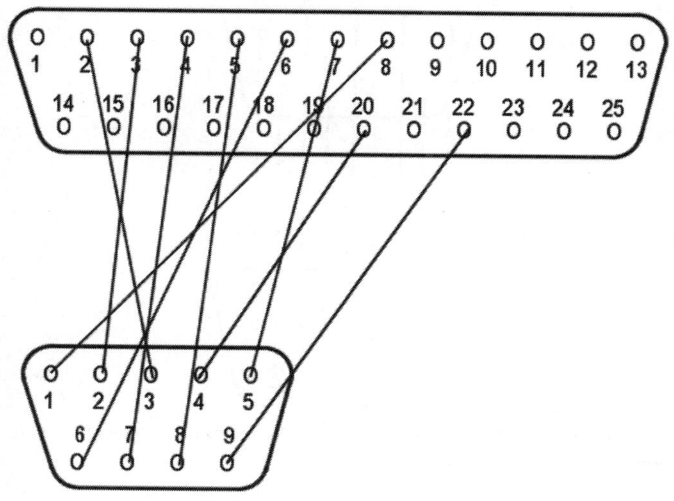

Figure 8.8 9-pin DTE to 25-pin DCE cable connection

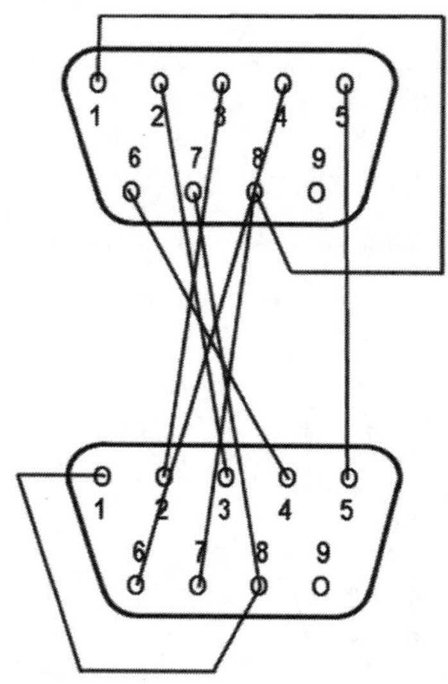

Figure 8.9 9-pin DTE to 9-pin DTE null modem cable connection

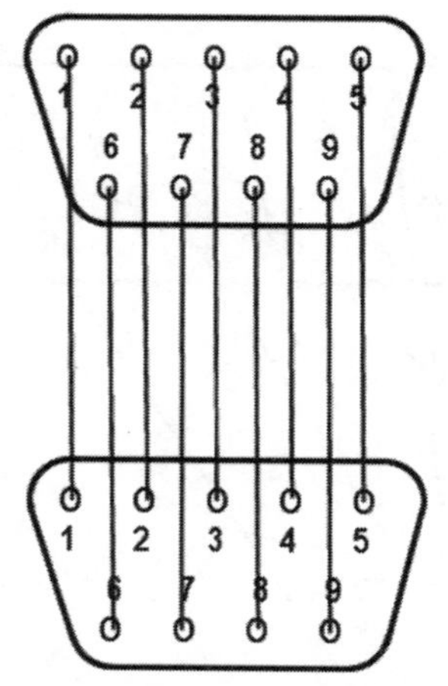

Figure 8.10 Straight through 9-pin to 9-pin cable. Can be used as a universal serial extension cable

1200 baud). When a successful link has been established, more complex protocols may be attempted in order to increase the data transfer rate or improve upon data checking.

Table 8.1 Loopback connections

Signals	**Pins linked (9 pin)**	**Pins linked (25 pin)**
RX and TX	2 to 3	3 to 2
RTS and CTS	7 to 8	4 to 5
DTR and DSR	4 to 6	20 to 6

In some legacy devices you may come across a bugged chip. Serial ports are driven by a UART and the early ones had the type number 8250. These chips have several race conditions and other flaws. The updated chips, the 8250A, 16450 or 16550, do not have these flaws. The 16550 is the most common. If the system you are troubleshooting has an 8250 chip, that is the trouble! They will work but only slowly.

9 Replaceable disk drives

Originally, floppy disks provided a means of exchanging data between computers, installing software on the hard disk or even backing up the data stored on your hard disk. They have been largely superseded by faster devices but there is still a demand for their use. PCs without floppy drives are still not common except in corporate networks.

This chapter begins by introducing the most commonly used floppy disk formats and the structure of boot records and file allocation tables (FAT). It also describes the floppy disk interface and the functions of the PC's floppy disk controller (FDC). The chapter also explains how to remove and replace a floppy disk drive.

9.1 Floppy disk formats

A variety of different floppy disk formats are supported by PCs but most are of historical interest only. Most if not all floppy disks are now the 1.44 MB format 3.25″ disks. They have two small square holes unlike the older 720 KB format which had just one. You may see markings such as DSDD or DSHD that refer to double sided double density or high density.

When a disk is formatted, the operating system writes a magnetic pattern on the surface of the disk. The pattern normally comprises 80 concentric 'tracks', each of which is divided into a number of 'sectors'. The magnetic pattern is repeated on both sides of the disk (note that some obsolete disk formats are designed for 'single-sided' disks).

> **TIP:** Floppy disks sometimes become contaminated with surface films (e.g. due to liquid spills or dropping it in your coffee). If you need to recover the data stored on such a disk, you should first carefully remove the disk by prizing apart its plastic housing, and then rinse it under a tap with warm running water. Do not use cleansers or detergents. When the surface of the disk has been thoroughly rinsed, you should dry the disk in warm (but not hot) air before replacing it in its plastic housing. During the cleaning process it is important to hold the disk by the edges or by the central hub ring. You should avoid touching the surface of the

disk. Once the disk has been reassembled, you should immediately attempt to copy the data to another disk or to a temporary directory on the hard drive and then discard the disk.

TIP: During disk formatting, you may occasionally notice that the drive spends some considerable time towards the end of the process (the heads may appear to be stepping erratically). This is a sure sign the operating system has discovered some 'bad sectors' on the surface of your disk. If the problem is severe, the formatting will be aborted (this also happens if track 00 is found to be bad). However, if it is not severe, the format will eventually be completed but the disk may not deliver the capacity that you expect. In such a case, it is worth formatting the disk a second time. If that still gives bad sectors, discard the disk.

9.2 The boot record

Floppy disks can be formatted as 'bootable' or data only. A floppy disk's 'boot record' (or 'master boot record') occupies the very first sector of a disk. The boot record contains a number of useful parameters and may contain code which will load *and* run (i.e. 'boot'). Bootable disks must also contain two other programs: IBMIO.COM and IBMDOS.COM (or IO.SYS and MSDOS.SYS). These, in turn, are responsible for locating, loading and running the command interpreter (COMMAND.COM). The parameters contained in the boot record include details of the disk format (e.g. the number of bytes per sector and the number of sectors per cluster).

Floppy disks can also be 'quick formatted'. All this does is to overwrite the boot record and FAT (file allocation table), leaving all the data still present. This usually gives no trouble but is not secure if the data is in anyway private.

9.3 Booting the system

When a PC performs a 'warm boot' or 'cold boot' (using the <CTRL> <ALT> <DEL> keys or by pressing the 'reset' button respectively), the ROM BIOS code initializes the system and then attempts to read the boot sector of the floppy disk in drive A:. If no disk is present in drive A:, the ROM BIOS reads the first sector of the hard drive, C:. Note that many BIOS setup programs actually give you the option of ignoring any operating system code that may be present on a disk placed in drive A: at bootup time. In such cases, the system will boot from the hard drive (i.e. drive C:) or a CD-ROM. Disabling the drive A: boot facility has the advantage that any disk containing a boot sector virus will be ignored. Unfortunately, there is also a downside to this – having disabled booting from drive A: you will have problems when drive C: eventually fails and the system refuses to boot because it cannot locate the operating system files! However, if (or when) this eventually happens there is no need for panic as all you need do is enter the BIOS setup routine to once again enable booting from drive A:.

9.4 Troubleshooting the floppy disk drive

Troubleshooting disk drives can be a complex task. In addition, it must be recognized that the drive contains highly sophisticated electronic and mechanical components which require both careful and sympathetic handling. Hence it is recommended that, at least for the inexperienced reader, consideration be given to replacing drives, this being more cost effective in the long run. In any event, fault diagnosis within drives should only be carried out when one is certain that the disk interface and controller can be absolved from blame. Thus, whenever the drive in a single-drive system is suspect, it should first be replaced by a unit that is known to be good.

TIP: Modern 3½″ disk drives are so inexpensive that it is not usually cost effective to carry out any repairs or head adjustments on them and many floppy disk drives are discarded when they first begin to cause problems. That said, the most frequent cause of problems is simply an accumulation of dust, dirt and oxide on the read/write heads. Thorough cleaning is all that is required to put this right!

TIP: The head alignment often varies from machine to machine. Disks written on one machine may be reported as faulty on another. Sometimes a solution is to perform a complete reformat of the disk on the second machine, copying the data over to the first machine.

The read/write heads of disk units require regular cleaning to ensure trouble-free operation. In use, the disk surface is prone to environmental contaminants such as smoke, airborne dust, oils and fingerprints, and these can be transferred to the read/write heads along with oxide particles from the coating of the disk itself.

Periodic cleaning is thus essential and, although this can easily be carried out by untrained personnel using one of several excellent head cleaning kits currently available, head cleaning is rarely given the priority it deserves. Thus, whenever a PC is being overhauled, routine cleaning of the heads may be instrumental in helping to avoid future problems.

> **TIP:** The read/write heads of a floppy disk drive are permanently in contact with the disk surface when the disk is in use. The heads can thus easily become contaminated with particles of dust and magnetic oxide. You can avoid this problem by cleaning your read/write heads regularly using a proprietary head cleaning disk and cleaning fluid. As a rough guide, you should clean the heads every month if you use your system for two or more hours each day.

> **TIP:** Some problems that look like trouble with a floppy are in fact caused by a virus. Some even quite old viruses cause trouble with the floppy boot sector and can be quite tricky to eradicate.

9.5 Replacing a disk drive

The procedure for removing and replacing a floppy disk drive is quite straightforward. You should adopt the following procedure:

1. Power-down and gain access to the interior of the system unit (as described in Chapter 2).
2. Locate the drive in question and remove the disk drive power and floppy disk bus connectors from the rear of the drive.
3. Remove the retaining screws from the sides of the drive (four screws are usually fitted).
4. Once the drive chassis is free, it can be gently withdrawn from the system unit. Any metal screening can now be removed in order to permit inspection. The majority of the drive electronics (read/write amplifiers, bus buffers and drivers) normally occupies a single PCB on one side of the drive.
5. The head load solenoid, head assembly and mechanical parts should now be clearly visible and can be inspected for signs of damage or wear. Before reassembly, the heads should be thoroughly inspected and cleaned using a cotton bud and proprietary alcohol-based cleaning solvent.

6. Reassemble the system (replacing the drive, if necessary) and ensure that the disk bus and power cables are correctly connected before restoring power to the system.

TIP: Take special care when replacing machined screws which locate directly with the diecast chassis of a disk drive. These screws can sometimes become cross-threaded in the relatively soft diecast material used to manufacture the exterior chassis of some drives. If you are fitting a new drive, you *must* also ensure that you use screws of the correct length. A screw that is too long can sometimes foul the PCB mounted components.

TIP: A 34-way male PCB header is usually fitted to $3\frac{1}{2}''$ drives. Unfortunately, the matching female IDC connector can easily be attached the wrong way round. You should thus check that the connector has been aligned correctly when replacing or adding a disk drive to your system. Pin-1 (and/or pin-34) is usually clearly marked on the PCB. You should also notice a stripe along one edge of the ribbon cable. This stripe must be aligned with pin-1 on the connector.

9.6 Zip and Jaz disk drives

Iomega's popular Zip disk standard provides a means of storing large amounts of data in a removable disk which is only slightly larger than a conventional $3\frac{1}{2}''$ disk. A Zip disk drive can be added internally (in a vacant drive bay) or externally (different versions are available for connection to a parallel or a SCSI connector). Internal drives are connected to the EIDE or SCSI interface in just the same way as modern floppy or CD-ROM drives. External drives are connected via a cable to the USB port.

9.7 Troubleshooting Zip disks

Some Zip disks give rise to the 'click of death'. These Zip disks use data on the disk to maintain tracking if the data is corrupted in any way, the read/write head searches in vain for the data. This in turn gives rise to 'hunting', where the read/write head moves back and forth but never finds the right data. It is this hunting that makes a clicking sound. Unfortunately, this problem occurred on the original drives and rendered the disk unusable. It would often herald the demise of the drive as well.

Computer hangs while it is reading or writing:

- Check power and cable connections.
- Check for a resource conflict. In Windows Control Panel, open the Device Manager. Devices that have conflicts with other devices will be shown with a yellow ! symbol. You will need to refer to the specific technical documentation for the device that is conflicting. These are usually available on the Internet. If you do have a conflict, Windows allows you to change the settings but will warn you that other devices may not work!
- If your drive has a SCSI interface, ensure you have a unique SCSI device number set for each SCSI device on the SCSI bus. Also, the first and last device on the SCSI chain must be terminated.
- Try the drive on a different computer.

TIP: To find specific device conflict information, use www.google.com and type the word 'conflict' followed by the names of the conflicting devices. Others have been there before you and you should find all you need to know.

10 Hard disk drives

10.1 Hard drive basics

The hard drive in a PC employs a rotating disk or disks. These disks are coated with a material that has certain magnetic properties that are persistent, i.e. once changed they stay that way for a long period until changed again. This is all that is required to store data as 1s and 0s.

The precise details of how the various types of hard drive actually work will be ignored, the most important points to note are:

- Capacity.
- Performance.
- Cost.
- Reliability.
- Compatibility with other systems/components.

Although we shall ignore the fine detail, some understanding of how a drive works is required to properly understand factors in the performance of a drive. The disk rotates at a constant speed and a *read/write head* is moved to nearly any point over the surface. If fed with the correct signals, this read/write head is able to affect the magnetic coating of the disk to store a binary 1 or a binary 0 or of course to read 1s and 0s from the disk. Clearly the speed at which this read/write head can be moved over the disk surface and the speed at which it rotates will have an affect on the speed of operation. The read/write head is usually

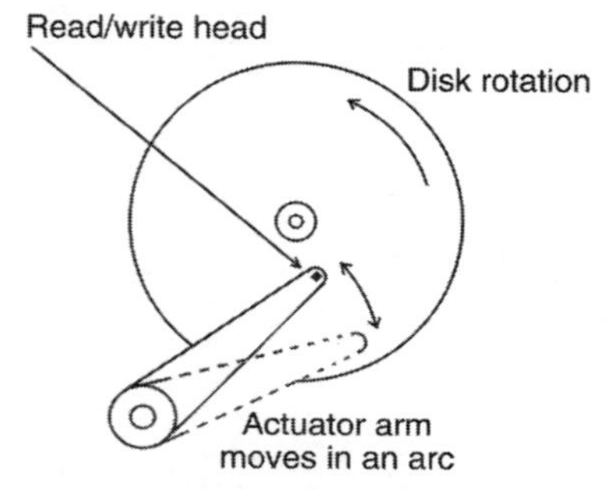

Figure 10.1 Layout of simple disk drive

is fine except that anyone can modify the 'standard' and still claim they are conforming. This leads to an ever growing list of 'standard' names which can and often does lead to confusion. This is especially true with disk drive interfaces.

10.3.1 IDE/ATA interface

You will see many adverts for *IDE drives*, a term that was originally used to differentiate the old drives that used a separate controller board from the (then) new *Integrated Drive Electronics* (IDE) drives. The trouble is, other devices also have 'integral' electronics so the term is of little use. Drives that are called IDE are really using the 'AT Attachment' standard so should be called ATA but most adverts use the term IDE. IDE/ATA is by far the most popular interface commonly available but variations are also known as ATA, EIDE, ATA-2, Fast ATA or Ultra ATA. The ATA Packet Interface or ATAPI is an extension that allows connection to CD-ROM drives, etc.

10.3.2 Original ATA

The original IDE/ATA supported two hard disks that worked with 'programmable I/O' or PIO modes and DMA (see section 10.3.8 on DMA).

10.3.3 ATA-2

With increasing performance demands the next standard, ATA-2, was brought out to support faster PIO, LBA and several other enhancements that need not concern us here. Ignoring a few minor details, ATA-2 is the same as Fast ATA, Fast ATA-2, or Enhanced IDE (EIDE). The differences are those between different manufacturers and is an example of a 'standard' being applied differently!

10.3.4 ATA-3

ATA-3 is an improvement to ATA-2 that introduces self-monitoring analysis and reporting technology (SMART). Although often confused with it, ATA-3 is not the same as Ultra ATA.

10.3.5 Ultra ATA

Ultra ATA (also called F, ATA-33, and DMA-33, etc.) is an even faster interface that uses a 33.3 M byte/s transfer mode using DMA. Ultra ATA is backwards compatible with previous ATA interfaces, so if you fit an Ultra ATA drive to a board that does not support the faster mode, it will still work at the slower speed.

10.3.6 ATAPI

The original ATA would not support CD-ROM drivers or floppy disks so the *ATA Packet Interface* (ATAPI) was introduced to bring the advantage of one standard to cover all the common drives. To make this work, an ATAPI driver must be installed; this is a piece of software called a 'device driver', and it communicates with devices using 'packets' of data. This is because CD-ROMs are quite unlike hard drives in their method of working so software makes up the difference.

10.3.7 PIO

IDE/ATA drives support both PIO and DMA transfer modes. Programmed I/O (PIO) is performed by the main CPU, it requires no support software but ties up the CPU when I/O is in progress. There are several PIO 'modes', each newer than the last, the fastest of which is supported by the latest IDE/ATA drives. As a performance comparison these modes give maximum transfer rates of

PIO mode 0	3.3 M byte/s
PIO mode 1	5.2 M byte/s
PIO mode 2	8.3 M byte/s
PIO mode 3	11.1 M byte/s
PIO mode 4	16.6 M byte/s

10.3.8 DMA

Direct memory access or DMA achieves data transfer without the CPU, either using the old (and rather limited) DMA chips fitted as standard to all PCs or using bus mastering of the PCI bus.

DMA mode

Single word 0	2.1 M byte/s (no longer used)
Single word 1	4.2 M byte/s (no longer used)
Single word 2	8.3 M byte/s (no longer used)
Multiword 0	4.2 M byte/s
Multiword 1	13.3 M byte/s
Multiword 2	16.6 M byte/s
Multiword 3	33.3 M byte/s (DMA-33)

Do not think that PIO mode 4 and DMA Multiword 2 will give the same overall performance. Both achieve 16.6 M byte/s but the PIO mode ties up the CPU whereas the DMA mode allows the CPU to

may be incompatibility between the two drives. If this is so, there is not likely to be a way to fix it.

10.7.6 Check power

Is the drive getting power? You can usually hear the disk spin up to speed. Check the power leads and swap them around with other devices known to work.

10.7.7 Configuration settings

If the computer locks up when HDD is installed and if it boots OK from floppy, type FDISK/MBR at the DOS prompt. This rewrites the MBR on the disk (master boot record). If FDISK.EXE is not on the floppy, it can be found on a Windows rescue disk.

If the drive seems to work when the PC has been booted from a floppy, it is possible the hard drive is:

- Not partitioned
- Not formatted
- Not bootable

(or even all 3!)

10.7.8 Partitioning

If the hard drive is not formatting to its design capacity:

- Check CMOS settings and verify cylinders, heads and sectors.
- Check that LBA (logical block addressing) is set for drives larger than 528 MB.
- Use FDISK to show partition information. Ensure drive is at correct capacity and is showing 100 per cent usage. If you have trouble with FDISK, delete all partitions and start again. Better still, use Partition Magic from Powerquest. See www.powerquest.com/partitionmagic

TIP: FDISK reports wrong size when using drives larger than 64 GB. According to Microsoft Knowledge Base article Q263044, 'When you use FDISK.EXE to partition a hard disk that is larger than 64 GB (64 gigabytes, or 68 719 476 736 bytes) in size, FDISK does not report the correct size of the hard disk.

The size that FDISK reports is the full size of the hard disk minus 64 GB. For example, if the physical drive is 70.3 GB (75 484 122 112 bytes) in size, FDISK reports the drive as being 6.3 GB (6 764 579 840 bytes) in size.'

10.7.9 Formatting

If you get a message such as 'cannot read sectors' or 'retry, abort, ignore' error while reading the drive, this indicates there are some bad spots on the drive. To fix these, either reformat the hard drive (and hence reinstall all the software) or use third party software such as Gibson Research's Spinrite. See grc.com/spinrite.htm

To format or reformat the drive, use FORMAT C: /s where the /s parameter means copy the system files in place from sector 0, i.e. make the drive bootable.

If you are sure the drive has been formatted, you can try SYS C: which will copy the system to the hard drive.

TIP: Do not *always* believe the manufacturers. On one brand new second drive fitted by the author, the bus speed setting, using the maker's software, was set to the recommended speed of 66. The drive appeared to work. After a few days, the machine user reported that 'the drive has deleted or renamed all my folders, I have lost my work'. The trouble was tracked down to the bus speed. Once set at 33 (the older standard) the drive worked, and continues to work without a hitch. As a bonus, all the folders also reappeared, along with his work! The maker's claimed the higher speed as a sales point.

10.7.10 Checklist

For a hard drive to work you must have:

- The BIOS settings correct.
- Ribbon cables plugged in with correct polarity.
- Power cables connected with correct polarity.
- The partitions must be correct (with LBA set on) using FDISK or Partition Magic.
- The drive must be formatted.
- The drive must be bootable (if it is the main drive).

10.8 Installing, replacing, upgrading a hard disk drive

The following procedure is recommended for installing or replacing a hard disk:

1. Power-down and gain access to the interior of the system unit.

2. Remove the hard disk's power and data connectors from the rear of the drive.
3. Remove the retaining screws from the sides of the drive (four screws are usually fitted).
4. Once the drive chassis is free, it can be gently withdrawn from the system unit (you may have to move cables around to clear enough space at the rear of the drive).
5. Verify the jumper configurations at the rear of the new drive. In particular, check the master/slave jumper settings (a master setting will be required if the hard disk is to be the bootable drive – the slave is not bootable and would normally be the second drive in a system fitted with two drives). If you are fitting the new drive as a second drive (not as a replacement for the first drive) you will also have to ensure that the original drive has its jumpers configured for operation as a master.
6. Fit the new drive and ensure that the data and power cables are correctly connected before restoring power to the system.
7. Apply power to the system and, when the memory check has been completed, enter the CMOS setup routine (on most systems you need only press the Delete key to do this). If the system BIOS supports a user programmable drive type you can enter the default parameters of the new drive (you will find the information in the handbook supplied with the drive or printed on the drive's label). The BIOS should offer you the option of automatically detecting the drive's parameters. This is usually quicker and more reliable than attempting to enter the information manually. Once the parameters have been accepted you should ensure that the new data is written to the CMOS memory before you exit from the setup program (the drive's parameters will not be stored and your drive will not be recognized if you fail to save this information).
8. Boot the system from a DOS/Windows disk boot disk that contains FDISK.EXE and FORMAT.COM. If you are installing a second hard drive you can boot from the existing C: drive (provided that it has been configured as a master and then use the FDISK.EXE and FORMAT.COM utilities from the existing hard disk.
9. Use the setup software that came with the disk. If the disk did not come with its own setup software, you must use the FDISK utility to set up the partitions and prepare the disk for DOS. Then you must use the FORMAT command to prepare the disk for data. The procedure is slightly different depending upon whether the hard disk is the only disk present or whether it is the second hard disk in a system with two hard disk drives.

The following stages are required if the hard disk is to be *the only drive* in a system (in which case DOS will identify the drive as drive C:):

10. You will have booted from drive A: (C: will not be recognized at this stage). At the A:> prompt, type FDISK followed by <Enter>. At the menu options, select option 1 to create a DOS partition. A second menu will now appear. From the new set of options select 1 to create a primary DOS partition. Select 'Yes' to make one large partition and it will automatically become active. Then quit out of FDISK.
11. At the A:> prompt, type FORMAT C: /S followed by <Enter>. This command will perform a high-level format on the drive and it will also transfer the system files in order to make the drive bootable. (Note that all modern IDE and EIDE drives are low-level formatted by the manufacturer and they *only* require a high-level format.)
12. When the format has been completed the new hard disk drive is ready for use. Remove the DOS disk from drive A: and press the reset button. If all is well, the computer should perform its full boot routine and you should be rewarded with a C:> prompt when the process completes. If not, check each of the stages from 5 to 11.

The following stages are required if the hard disk is to be the *second drive* in a system (in which case DOS will identify the drive as drive D:):

1. You may have booted from drive A: or from drive C: (D: will not be recognized at this stage). At the A:> (or C:>) prompt (assuming that FDISK.EXE and FORMAT.COM are located in a directory listed in the PATH statement), type FDISK followed by <Enter>. At the menu options, select option 5 to switch to the second then enter Fixed Disk Drive Number 2. Next choose option 1 to create a DOS partition, then select option 1 again to create a primary DOS partition or option 2 to create an extended DOS partition. Then quit out of FDISK.
2. At the DOS prompt, type FORMAT D: followed by <Enter>. This command will perform a high-level format on the drive and it will also transfer the system files in order to make the drive bootable. (Note that all modern IDE and EIDE drives are low-level formatted by the manufacturer and they *only* require a high-level format.)

TIP: One of the most common problems with hard disks is related to temperature. If you use your PC in an unheated room on a cold morning you may find that the hard disk does not operate correctly. You may also encounter similar problems when it is very hot. Problems can often be reduced by ensuring that the hard disk is formatted at the normal working temperature of the machine. In other words, before you attempt to carry out a hard disk format, you should wait for between 15 and 20 minutes for the machine's internal temperature to stabilize. Never attempt a format first thing on a cold and frosty morning or last thing in the afternoon where the temperature has been building up all day. The difference in these two extremes can be greater than 20°C and, like you, your hard disk may well not perform consistently over the whole of this range!

TIP: Never turn off your computer when it is performing a hard disk access (you should *always* check the hard disk indicator before switching off). Failure to observe this simple rule can cause many hours of frustration and may even require drastic action on your part to recover lost or corrupt data. If you are unlucky enough to be presented with a system that boots up with an error message relating to the hard disk or tells you that there is no space available on the disk or that directories are missing or corrupt, it is worth trying to recall what happened when you *last used* the system. If you do suspect that something was not right when the PC was switched off (e.g. the user switched off while Windows was manipulating a giant swap file) it is worth running the ScanDisk utility.

TIP: If your hard disk fails, don't immediately rush to reformat the disk. There are several things that you can do before you resort to this course of action. If Windows does not recognize the existence of your drive you should use the BIOS setup routine and check the CMOS RAM settings. If Windows recognizes your drive but you are not able to either write to or read from it, you should use the ScanDisk utility to gain some insight into the structure of the disk (or at least how Windows currently perceives it).

TIP: Windows 95/98 give you an option to create a boot disk. This takes much of the effort out of having to perform this task manually. Place a blank, formatted disk in the floppy drive and select 'Settings', 'Control Panel' and 'Add/Remove Programs'. Then select 'Startup Disk' tab and click on 'Create Disk'. Windows will then copy the necessary files to the disk (starting with

COMMAND.COM). If necessary, Windows 95 will prompt you to insert the original Windows 95 disk from which the required file can be obtained. Alternatively, Windows will ask you to specify an alternative location where the required file can be found. Once the process has been completed, you will have a floppy disk that can be used to boot the system when and if the hard disk fails.

TIP: You can improve file access times by using a 'disk optimizer' or 'defragmenter' utility. In Windows, the Defrag utility will perform the task. If Defrag detects errors in the file structure on the hard disk, it will halt and invite you to run ScanDisk. ScanDisk will check the file structure on your disk and can automatically fix any errors that it encounters. ScanDisk is usually run before Defrag when performing a routine check on a hard disk drive. Both ScanDisk and Defrag can usually be found in Programs → Accessories → System Tools. Defragmenting a disk is only useful if it has been used a great deal, i.e. if much data has been written then deleted leaving gaps.

TIP: Windows 98 takes the art of disk optimization one stage further with its 'Defragmenter Optimization Wizard'. This utility tracks the programs that you run most often, then clusters these programs on the fastest part of your hard disk. To use the utility, you must close down all applications before launching the wizard. Once activated, the wizard asks you to specify the programs that you run most frequently. It then launches these programs and notes which files are accessed before moving them to their new location on the hard drive. Grouping files together minimizes the seek time and ensures that your applications launch and run in the fastest possible time.

TIP: Defrag and ScanDisk sometimes fail to run, they report 'restarting' before finishing. To avoid this, either:

1. Stop all processes except 'Explorer' by using CTRL-ALT-DEL then 'end task' for each process, then use defrag or scandisk

or

2. Reboot the machine in safe mode

With either method, ensure printers and other external devices are turned off at the power source as they are often responsible for writing to the hard disk, causing Defrag/ScanDisk to restart.

> **TIP:** Take special care when replacing the screws used for mounting a hard disk drive. It is *very* important to use the correct length of screw. Internal damage can easily be caused by screws that are too long!

10.9 Recovering from disaster

Not only is it important to have a backup strategy but it is also important to be fully aware of the implications of that strategy. The secret of having an effective backup strategy is being able to accurately assess the risks that you might be subject to. Remember, too, that you are not just protecting against hard disk failure – other disasters can also render your data inaccessible!

10.9.1 Risk assessment

Consider each of the following possible scenarios:

(a) Your hard disk suddenly fails and all the data stored on it becomes inaccessible.
(b) Your laptop computer is stolen and the police are unable to recover it.
(c) Your office, and all its contents, are destroyed by fire.
(d) You discover that a virus has corrupted most of the files on your hard disk drive.

In case (a) you would be adequately protected by a recent backup kept locally. Simply replace the hard disk drive and restore its contents from the backup device. Likewise in case (b). Case (c) is rather different – you will only be able to recover from this type of disaster if you keep a remote backup (or have a totally fire-proof safe!). Case (d) requires an effective virus recovery program – simply restoring your files from your most recent backup may only serve to reinfect your system!

Scenarios (a) to (d) are all examples of risks that you might have to face one day. There may be others that apply to your own individual situation. Only *you* can make a realistic assessment of these risks and how likely they are to occur!

10.9.2 Backup strategy

A backup strategy involves answering the following questions:

- 'What to backup?'
- 'When to back it up?'

- 'Where to back it up?'
- 'How to back it up?'

'What to backup?' could include:

- All files (including hidden and system files).
- All data files.
- All files in the My Documents directory.
- All files created or modified since a given date.
- All files modified or added since the last backup (i.e. an *incremental* backup).

'When to back it up?' could include:

- At the end of each working day.
- At the end of each working week.
- On the first and third Fridays of each month.
- At the end of each month.
- Overnight, after close of business each Saturday.
- After each new report has been completed.

'Where to back it up?' could include:

- To a set of floppy disks kept in the bottom drawer.
- To a QIC tape stored in the IT department's fire safe.
- To a Zip disk kept at the branch office.
- To a network server in the company's computer centre.
- To an FTP Internet site.

'How to back it up?' could include:

- Using a single floppy disk drive and Winzip or PKzip to combine and compress files.
- Using floppy disks in drive C: and the commercial backup utilities.
- Using an internal Zip disk drive and Iomega's proprietary Zip tools.
- Using an external Jaz tape drive and Iomega's proprietary Jaz tools.
- Using an integral QIC tape drive and Central Point's backup.
- Using one or more writable CD-ROMs.
- Using a backup directory on a local area network.
- Using an external hard disk drive and Windows File Manager.
- Using a Travan or DAT external SCSI tape drive.
- Using Netware to upload compressed files to an Internet site.

TIP: An effective backup strategy for most users is to perform an incremental backup on a regular basis (daily or weekly) with a full backup (i.e. all files) performed on a less regular basis (weekly, fortnightly or monthly). It might also be worth considering the use of different types of media for these two operations (e.g. Zip disks for incremental backup and Travan or DAT tape for full backups). If you then store at least one of these backups off-site (together with an earlier 'grandfather' or 'father' copy of the backup made using the *other* media) you will have a very high degree of protection. If you do adopt a mixed strategy like this it is, of course, essential to label all of your backups so that you know which one is the most recent!

TIP: It can take some time to perform a full backup during which your system may be unavailable for normal use. If this is a problem you should consider automating your backup so that it will be performed when you are out of the office (most good backup software will allow you to do this). Alternatively, you should make time in your working routine so that your system is able to perform the required backup operation on a regular basis.

11 Displays

The PC supports a wide variety of different types of display. This chapter explains the most commonly used display standards and video modes. It also tells you how to get and set the current video mode and provides some basic information on how the PC produces a colour display.

11.1 PC display standards

The video capability of a PC will depend not only upon the display used but also upon the type of 'graphics adapter' fitted. Most PCs will operate in a number of video modes which can be selected from DOS or from within an application.

The earliest PC display standards were those associated with the monochrome display adapter (MDA) and colour graphics adapter (CGA). Both of these standards are now obsolete although they are both emulated in a number of laptop PCs that use LCD displays.

MDA and CGA were followed by a number of other much enhanced graphics standards. These include enhanced graphics adapter (EGA), multi-colour graphics array (MCGA), video graphics array (VGA), and the 8514 standard used on IBM PS/2 machines.

The EGA standard was followed by VGA and now SVGA ('super VGA'). The first generation of VGA displays (1987) were based on 8-bit controllers and only supported a resolution of 640 × 480 pixels. These were followed (in 1989) by second generation controllers and displays capable of a resolution of 800 × 600 pixels. At the same time, architecture moved from 8 bits to 16 bits.

The third generation of SVGA controllers and displays appeared in 1991. These systems supported display resolutions of up to 1024 × 768 pixels. At the same time, video cache memories became commonplace together with the VL bus interface which provides a vast increase in display speed and overall performance.

Today's SVGA controllers – fourth generation controllers – provide even more acceleration with wider video ports and much larger display memories (e.g. 8 M byte). In addition '3-D' shading and texturing is provided by many controllers to enhance multimedia and games programs. A very high degree of integration is now provided in the controller chipsets which move graphic data around in 32- or 64-bit chunks.

11.1.1 Pixels

When a graphical image is shown on a computer screen, it is made up from a large collection of dots. Each of these dots is called a *pixel*, a word which is short for *picture element*. Pixels are arranged in rows and columns; typical numbers of rows and columns are shown in the table below.

Name	Columns	Rows
VGA	640	480
SVGA	800	600
	1024	768
	1280	1024
	1600	1200

The number of pixels per screen is known as the resolution. The higher the number of pixels the better the resolution or the finer the detail that can be shown.

Each pixel may have just one colour at a time, chosen from a set of colours. Suppose each pixel could be just one from a selection of 256 different colours. This would mean that a number must be assigned to that pixel. If bright red was colour number 37, then that one pixel would be stored as the value 37. Since 1 byte is made up of 8 bits, the largest number that can be stored in 8 bits is 255. If you include the value 0 then it is possible to store one of 256 different colour values in 1 byte, or the pixel can be one of 256 different colours. Another way to state this is that $2^8 = 256$.

If you use less memory than 1 byte per pixel, say 4 bits per pixel (half a byte), then you must be content with fewer colours. With 4 bits, the largest number you can store is 15, so (including the value 0) the number of possible colours is 16 or 2^4. (Half a byte is called a nibble!)

More generally, the number of colours that are available is given by 2^N. If you choose to use 8 bits per pixel then the number of colours is $2^8 = 256$, or with 24 bits per pixel, the number of colours is $2^{24} = 16.7$ million.

11.2 Video graphics

Most video graphics systems use a value of N that divides evenly into bytes, so values of 2, 4, 8, 16 or 24 are common, values such as 3, 5, 7 are not. The table below shows the number of colours available.

Bytes	Bits	Number of colours
$\frac{1}{4}$	2	$2^2 = 4$
$\frac{1}{2}$	4	$2^4 = 16$
1	8	$2^8 = 256$
2	16	$2^{16} = 65\,536$ (64k), Hicolor
3	24	$2^{24} = 16.7$ million, Trucolor

The term *colour planes* is sometimes used to describe the power of 2 so a $2^{24} = 16.7$ million colour setting would be described as a 24-bit colour plane. This comes from the design of the original VGA graphics card.

16-bit colour is called *Hicolor*, 24-bit colour is often known as *Trucolor* and is used where the better graphical image quality is required. Some scanners are now offering 30-bit colour although you could not realistically expect the full $2^{30} = 1\,073\,741\,824$ colours!

Consider some realistic limitations of human perception of image quality. If you have 16.7 million colours, can you see all of them? There are several answers to this.

1. Humans can perceive approximately 10 million colours.
2. The phosphors in the monitor cannot reproduce all the colours that you can see, a really convincing brown colour is very hard to make.
3. If you have a screen set to 1024*768 pixels, you have less than 16.7 million pixels. To have enough pixels to have one each of 16.7 million colours, you would need a resolution of 4730*3547 (maintaining the width to height ratio of 4:3).

The main reason to have 16.7 million colours is not to use them all but to have sufficient shades of each primary colour to reproduce a realistic shaded representation of an object.

11.3 Graphics cards

The original design of PCs had no facility to output graphics. The method used get around this problem was to have a separate *video card* or *graphics card* plugged into the main board. This card contained the video RAM and some ROM BIOS that contains the code required to write pixels, etc. More RAM allowed more colours or higher screen resolutions. Modern PCs may have the video card incorporated on the main board or as a separate component.

Modern video cards incorporate a CPU to speed up the graphics process. Imagine this problem: you wish to draw a single line at an angle on the screen and you know the colour and the start and end points of the line. Somehow the position of all the pixels that form the line must also be calculated. If this is done using the main CPU then that CPU is not available for calculating new graphics data so the system is relatively slow. If a CPU on the graphics card is dedicated to this task (known as *vector generation*), the main CPU is available for other calculations so speeding up the process greatly. A better enhancement is obtained when solid in-fill colours are required. A 100 by 100 pixel square needs $100 \times 100 = 10\,000$ pixels to be coloured. When this is done by a dedicated CPU, the main CPU has only to calculate the corner positions, just four points.

Note that non-vertical or non-horizontal lines cannot be quite smooth as they are made up of pixels, a phenomenon known as aliasing. The same is true for circles, etc. and extra pixels can be added in different colours to smooth out the line; this is called anti-aliasing. There are a number of algorithms used to calculate these extra pixels which require a fairly large CPU 'overhead' and if not done by the graphics card, would seriously slow down the main tasks of the host PC. A disadvantage of anti-aliasing is that lines, etc. become wider so reducing the apparent crispness of detail in some images.

Figure 11.1 Anti-aliasing

> **TIP:** As the resolution is increased, the size of individual pixels displayed, and therefore the 'step' between them, is reduced. As the pixel and step size is reduced, the jagged edges become less jagged and therefore less noticeable. For this reason, you should always use the highest resolution available on your system but note that this will depend upon the capability of your graphics controller *as well as* your display. To put this into perspective, it is worth comparing some of the most popular video standards for computers with those used in television.

Most PCs are now sold with at least SVGA or *super VGA* graphics cards giving at least 1024 by 768 screen resolution or better and at least 4 MB of video RAM. If you need more colours, then you have the option to increase the amount of video RAM. A video card set to 1024 by 768 Hicolor requires at least 2 MB, if you want this resolution and 24-bit Trucolor, you need 4 MB as shown below.

Video RAM (VRAM) is a specialized type of RAM that allows *dual porting*, i.e. it allows the CPU to access the RAM at the same time as the video circuits. If a screen refresh of, say, 85 Hz is used, the video system must access the RAM 85 times a second. VRAM allows the CPU access during the same time.

11.4 Video RAM required

A screen resolution of 800×600 pixels will yield $800 \times 600 = 480\,000$ pixels. If each pixel needs half a byte it will allow for 16 colours so the screen will require $480\,000 \times 0.5 = 240\,000$ bytes of storage.

$800 \times 600 \times 256$ colours require $(800 \times 600 \times 1) = 480\,000$ bytes of storage because each pixel will need 1 byte. $2^8 = 256$ different colour combinations. 480 000 is roughly half a megabyte of RAM.

In general, calculate the storage required for 1 pixel remembering that 2^N = number of colours where N = the number of bits required. Next multiply by the number of pixels on the screen.

Table 11.1 shows what RAM is required for single static screens. Animated graphics, texture mapping, vector generation, etc. all require more RAM. The result is that games PCs need large video RAM, office PCs generally do not.

TIP: When purchasing or upgrading a display and/or a graphics adapter there are a number of important questions that you should put to the supplier. These include:

- How many and what video standards and resolutions are supported?
- How much RAM is supplied on the adapter card? More RAM is good but only if you want animated graphics, games, etc. A PC used for office software will not benefit from a large video RAM.
- What type of RAM is supplied on the adapter card (high-speed VRAM and EDO RAM will result in faster video throughput)?
- What card format (AGP or PCI) is supplied?
- Do I need an MPEG compatible card (useful for multimedia video playback)?

Table 11.1 Video RAM required for a single static video screen

	Pixels			256 colours		64K colours (16 bit or Hicolor)		16.7 million colours (24 bit or Trucolor)	
	Horizontal	**Vertical**	**Total**	**bytes**	**Mbytes**	**bytes**	**Mbytes**	**bytes**	**Mbytes**
CGA	640	200	128 000	128 000 =	0.12	256 000 =	0.24	384 000 =	0.37
EGA	640	350	224 000	224 000 =	0.21	448 000 =	0.43	672 000 =	0.64
VGA	640	480	307 200	307 200 =	0.29	614 400 =	0.59	921 600 =	0.88
NTSC TV	672	525	352 800	352 800 =	0.34	705 600 =	0.67	1 058 400 =	1.01
PAL TV	767	575	441 025	441 025 =	0.42	882 050 =	0.84	1 323 075 =	1.26
SVGA	800	600	480 000	480 000 =	0.46	960 000 =	0.92	1 440 000 =	1.37
VHR	1280	1024	1 310 720	1 310 720 =	1.25	2 621 440 =	2.50	3 932 160 =	3.75
HDTV	1920	1080	2 073 600	2 073 600 =	1.98	4 147 200 =	3.96	6 220 800 =	5.93
UHR	2048	1536	3 145 728	3 145 728 =	3.00	6 291 456 =	6.00	9 437 184 =	9.00

TIP: When purchasing a display, it is important to check that it will cope with a range of different vertical and horizontal scanning frequencies. Most 'multi-sync' compatible monitors will operate with vertical scanning frequencies between 50 Hz and 100 Hz and horizontal scanning frequencies between 30 kHz and 38 kHz. If your chosen display cannot accept this range of scanning frequencies you will be unable to make use of the higher resolution text and graphics modes.

TIP: A screen saver will help to protect your system against the long-term effects of phosphor burn. It will not, however, eliminate the effects of constant power-on since the cathode ray tube heater will still be energized even when the screen saver is operational. Tube emission slowly deteriorates due to removal of the active material at the heater/cathode – a screen saver will not protect against this.

TIP: Certain screen savers have been known to crash a system when left for long periods. For this reason, you should always save your work before leaving the system and the screen saver to do its work. Indeed, there is no real reason to have a screen saver at all – just reducing the setting of the display's brightness control will have the same effect, at no cost and with less risk!

TIP: There has been some debate recently about whether it is better to leave a computer system (and its display) operational 24-hours a day rather than to switch it on and off (the act of switching on and off places additional electrical strain on a system). Calculations of mean-time-to-failure (MTTF) show that MTTF is reduced significantly if a system is left permanently on. Similarly, MTTF is reduced if a system is switched on and off several times each day (e.g. when taking breaks). The best compromise (and longest MTTF) can be obtained by switching on and off *once* each working day and leaving the system off at night and at weekends. In addition, when the user leaves the workstation he/she should either turn the brightness control down to a low setting or should ensure that a reliable screen saver becomes operational after a few minutes of non-use.

Another way to look at this problem is to consider the MTTF and the estimated time to obsolescence. Most commercial computer equipment is discarded in good working order!

TIP: Modern video cards and their drivers offer you a choice of colour definition and resolution. To adjust these values from within Windows you need to select Settings, then Control Panel, then Display and then choose the Settings tab. From here you can select the number of colours that will be displayed on the screen as well as the screen resolution (note that you may have to accept a trade-off between these two parameters depending upon the amount of graphics memory available). The colours settings available with most controllers include:

- 16 colours.
- 256 colours (selected from a palette of 262 144 colours).
- 'Hicolor' (16 bit – 32 768 on-screen colours).
- 'Trucolor' (24 bit – 16 777 216 on-screen colours).

Some experimentation will be required if you want to get the best out of your system. Much will depend upon the type of applications that you run. For example, 16 or 256 colours will be perfectly adequate for word processing and spreadsheet applications but can be woefully inadequate for graphics and digital photography.

If you do not have sufficient video RAM, you will not be able to get the highest number of colours at the highest resolution.

11.5 Display types

11.5.1 The CRT

The glass screen you see uses essentially the same image forming technique as domestic television, the image is made up of several hundred lines of a glowing substance called a 'phosphor'. This substance is made to glow by being 'hit' by a beam of electrons radiating from a point source at the back of the CRT. A fairly sophisticated arrangement of components in the CRT causes the electrons being emitted from the point source to be formed into a thin beam. If the beam were to be kept still, all you would see is a single point of light where the beam hits the phosphor on the inside of the front of the screen. Other components allow this beam to be moved anywhere on the screen, either horizontally or vertically.

A picture is made up by very quickly moving or scanning the beam from one side of the screen to the other in a series of lines that cover the whole of the visible portion of the screen. The detail in the picture is provided by changing the brightness of the fast moving beam. The phosphor has a property that causes it to glow for a little while longer after the beam has passed, this gives the illusion that the screen is evenly illuminated at all times.

11.5.2 Raster scan screens

To display a screen image on a CRT, an electron beam is focused on to the front of the screen; this screen is coated with a material that glows when struck by a stream of electrons. The beam is scanned from left to right to form a line across the screen, a scan line. Once each line is formed, the beam is made to return (or fly back) to the side of the screen and down one line, ready for the next line. Once the bottom of the screen is reached, the beam is moved to the top and the process repeated, each screenful of lines is called a *raster*. A picture is produced by changing the intensity of the beam as it traverses the screen and synchronizing this change with the intensity of the image required.

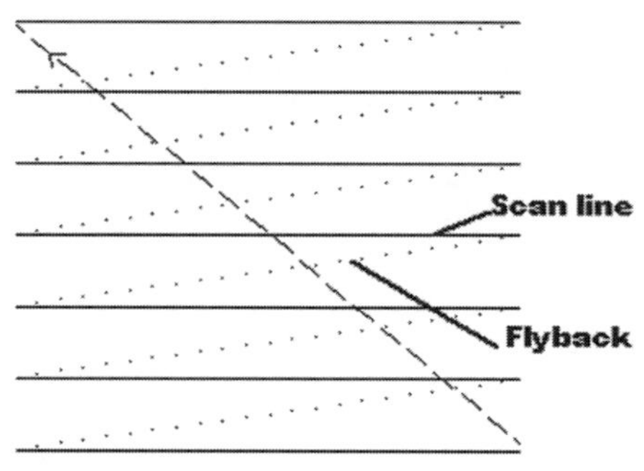

Figure 11.2 Raster scan

The number of times the raster is repeated per second is called the refresh rate. Slow refresh rates are seen by humans as flickering so rates of at least 72 times a second or 72 Hz are used to display a steady image, although 75 Hz is better. Normal TV screens use a raster scan but are slower than 72 Hz, a fact easily seen if you observe a TV screen in your peripheral vision where movement will be more obvious to you. This is even more obvious if you observe this at an electrical retailers shop where you have many screens at once. If they are arranged down the side of the shop and you look down the centre of the shop slightly away from the TVs, you will see a very marked flicker in your peripheral vision.

11.5.3 Colour screens

Colour CRT screens use three electron beams, one each for red, green and blue parts of the image.

An electron beam does not have a colour. To achieve colour, small areas of the screen are each allocated red, green and blue parts. These small areas are called slots and each of the three electron beams is made to hit exactly the right point, the 'red' beam hits a point that glows red, etc. The distance between each one is called the slot pitch or dot pitch. Generally, the smaller the slot pitch the sharper the image because there are more points of light to make up the image. Different shades of each colour are made up by varying the relative brightness of each of the red, green and blue parts of each slot.

Slots are only visible by using a powerful magnifying glass close to the screen. They can be observed on video monitors and domestic televisions and it can be seen that different makers use different shaped slots.

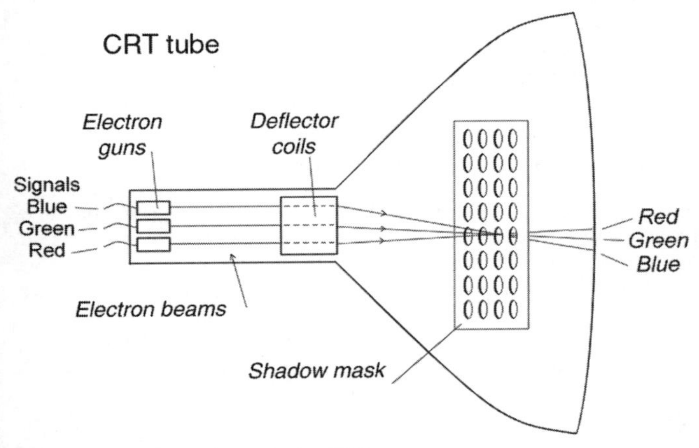

Figure 11.3 Basic layout of a CRT

Experiment: Use a magnifier to view a white portion of a monitor. You will see no white dots, only red, green and blue. All the colours you see are made of these three colours simply by mixing them in different amounts with different brightness. It can take some time to come to terms with the fact that there are no white portions of the screen. Most of your visual experience is concerned with your brain rather than your eyes, in this case your brain has 'manufactured' the white you see; your eyes are only receiving red, green and blue light.

Experiment: Observe the slots or dots carefully again with a powerful magnifier. Depending on the make of monitor you are using you will see dots or slots. Notice the slots have no effect on the image pixels, slots and pixels are completely different, pixels are a function of the computer graphics system and slots are fixed by the monitor maker. Move the image from side to side with the horizontal adjustments on the monitor. The slots stay still, the pixels move. You must not confuse slots with pixels; each pixel is a picture element and may take up different physical sizes on one computer by changing to different video modes. The number of slots available on one screen is fixed at time of manufacture of the screen. If you compare two video monitors with the same slot pitch but different sizes, the larger one will have more slots across the screen so will show each pixel more clearly. This is the reverse of the rule with domestic televisions, where larger screens show less sharp images.

11.5.4 Interlacing and refresh rates

On TV systems or old PCs, the beam cannot scan all the lines required to maintain a good number of full images per second, hence cannot provide an adequate refresh rate. To get around this, a system is used that scans only alternate lines, i.e. lines 1, 3, 5, 7, etc. to the bottom then the beam returns to the top and scans lines 2, 4, 6, etc. The result is half the number of full frames per second and is called interlacing; it is the system used on domestic televisions. This is highly undesirable on video monitors as the image can be seen to vibrate slightly. The human response is for the eye to try to keep up with the moving image, a process that leads to sensations of 'tired eyes' because your eye muscles must continuously move your eyes to adjust to the slightly moving image. Better video systems use non-interlaced screens that produce very steady images. Most people now agree that screen refresh rates of 75 Hz or better are required to achieve a good quality, steady image.

11.5.5 Laptop screens

There are many different kinds of LCD screen on the market, but they include:

- Passive supertwist nematic displays.
- Active-matrix displays.

11.5.6 Passive supertwist nematic displays

Passive supertwist nematic displays make use of a material, a nematic liquid crystal, that is able to change under the application of an electric field. The change in the nematic liquid crystal is in the way that light is polarized when passing through it.

In its normal state with no electric field applied, the polarization of the light that passes through it follows a twisted path. If the liquid crystal is put between two polarizing filters, each one with its axis of polarization at more than 90 degrees to the other, the twisted path of the liquid crystals causes any light from one side to be transmitted out through the other side. When a voltage is applied across the polarizing screens, the light in the liquid crystals no longer travels in a twisted path and so is blocked by one of the polarizing filters. This is used to turn pixels on and off.

11.5.7 Active-matrix displays

Active-matrix displays incorporate a transistor to control each pixel on the screen. These TFTs or thin film transistors are made into a matrix, each point in the matrix forms one pixel. Each TFT has opaque parts so not all the area assigned to a pixel is able to produce light. The ratio of the area of light production to the opaque part is called the aperture ratio and is taken as a measure of quality of the screen. A value of 30 per cent is good but a value of 50 per cent is better and this is achieved in the better screens.

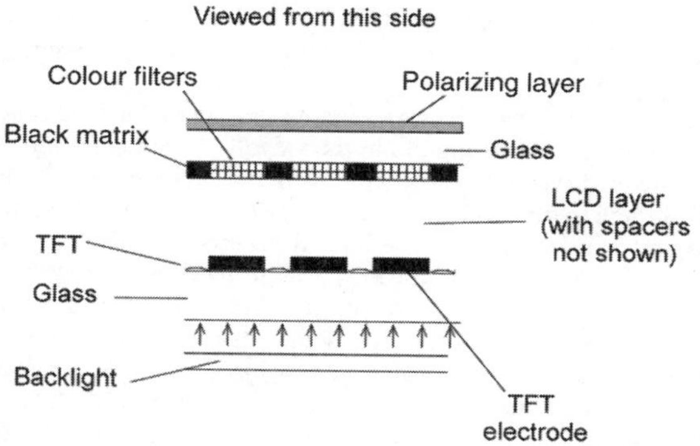

Figure 11.4 Simple diagram of active LCD screen

The backlight shown is a cold cathode fluorescent light source. The light from this is attenuated by the black matrix. The higher the aperture ratio, the brighter the image.

Compared with CRTs, the LCD screens found on notebook PCs are:

- Physically smaller.
- Have lower power consumption (allow batteries to be used).
- Lighter.
- Display good contrast with bright saturated colours.
- Show a crisp image.

There are problems when compared with CRTs – LCD screens have a:

- Viewing angle.
- Slow response time, they are not good for animated games, video, etc.
- High cost. The LCD screen accounts for most of the price difference between a desktop machine and a notebook PC. Active LCD displays are more expensive than the passive type but produce a much brighter display.

Web links for LCD screens:

Hewlett Packard: the first HP liquid crystal display:

www.hpmuseum.org/journals/hp41/41lcd.htm

Seiko LCD information:

www.seiko-usa-ecd.com

11.6 Troubleshooting displays

Most CRT sceeens are very reliable but problems can arise:

- If the display is blank and is used by children or students, check the brightness/contrast controls. Immature little fingers seem to like adjusting things!
- If the screen shows one colour missing (red, green or blue), the usual cause is a bent or missing pin in the video lead. Some monitors use a separate lead for each of the red, green and blue signals. Unplugging one of these in turn will cause the failure of that colour. If removing a lead makes no difference, you have found the problem!
- A flickering screen is often caused by an incorrect software setting. For some reason, a fresh installation of Windows 98 refuses to detect the correct video hardware, causing values of 800 by 600 at 256 colours to be set. The refresh rate may also be too slow, causing the flickering. Most humans find a refresh rate of at least 75 Hz will give a display that looks static. Reinstall the driver software for the monitor and graphics card.

- If the monitor seems to work but is too dull or too bright, use a piece of software to set the colours and 'gamma', the brightness and contrast combination. Adobe's Photoshop comes with gamma.exe and is an excellent program for setting the monitor.
- If the monitor shows odd colours on one side, check for the presence of a strong magnetic field near the monitor. The usual culprit is hi-fi speakers. The magnetic field pulls one or more of the electron beams to one side, causing them to become misaligned with the coloured phosphorescent dots on the screen.

> **Warning:** The voltages inside a CRT monitor are very high. On *no account* remove the cover or touch any internal component if the lead is still plugged into the power source. Repairing CRTs is a specialist task.

Laptop displays are very expensive and quite delicate:

- You may find that one or more pixels are always 'off'. Laptop manufacturers will replace a laptop if these are above a certain number, but one or two blank pixels is quite common.
- Replacing a laptop screen is not for the faint hearted as failure will be expensive. It is best to get it fixed by a specialist.

11.7 Troubleshooting video adapters

- Most trouble with video adapters is caused by software errors. To resolve the problem, reinstall the driver. If that fails, look on the maker's website for specific advice.
- If the operating system will not allow the screen resolution or number of colours that the adapter will allow, uninstall the driver software and reinstall. If this fails, obtain the latest driver from the manufacturer's website and try again.

12 Viruses

A virus is simply a program that has been designed to replicate itself in every system that it comes into contact with. It is more or less successful in this goal depending on several factors, including the sophistication of the virus, the level of anti-virus protection present and the 'habits' of the user.

Sooner or later all computer users have to come to terms with this particular nuisance. Don't think that it can't happen to you; just like hard disk failure it can and will happen to you sometime!

This chapter explains what a virus is and briefly describes the most common types of virus. It explains how your system can become infected by a virus as well as the simple steps that you can take to avoid infection. The chapter also describes the procedure for detecting and removing a virus from your system.

> **TIP:** When your hard disk fails to boot or your system locks up for no apparent reason, don't always jump to the conclusion that you have a virus. In most cases you won't have!

12.1 Types of virus

Opinions differ, but conservative estimates would suggest that there are currently many thousands of known viruses in existence and many of these exist in a number of different strains. Some viruses originated more than a decade ago but may appear with monotonous regularity in a variety of new disguises. In addition, completely new (and often highly sophisticated) viruses appear from time to time.

Viruses tend to fall into one of several main categories depending upon their mode of operation (i.e. whether they attach themselves to files, overwrite boot sectors, or pretend to be something they are not). The following main types exist.

12.1.1 Boot sector viruses

A boot sector virus copies itself to the boot sector of a hard disk or floppy disk. Boot sector viruses overwrite the original boot code and replace it with their own code. The virus thus becomes active every time the system is booted from an infected disk. The virus will then usually attempt to make copies of itself in the boot sectors of other disks.

12.1.2 Companion viruses

Companion viruses masquerade as an original program. They run before the program that acts as their companion – executing that program when they complete. This is sometimes referred to as a 'spawning' virus.

12.1.3 Software bombs

Software bombs are not really viruses. They are, however, a rather unpleasant form of malicious coding designed to cripple a program (or a system!) under a given set of circumstances (e.g. elapsed time, failure to enter a password). Unlike a virus, a software bomb does not attempt to replicate itself. It is simply embedded in the code to disable the software and/or the system on which it runs.

Software bombs can produce a variety of undesirable effects, some more unpleasant than others. The least problematic type simply erases itself from memory (together with any data that you were working on at the time). Others are liable to lock up your system (often until a cryptic password is entered) and some will delete programs and data files from your hard disk. The most extreme form of software bomb will effectively 'trash' your hard disk – yet another reason for making a regular backup!

12.1.4 Trojan horses

Trojan horses are destructive programs disguised as legitimate software. Sometimes these programs may purport to add some extra features to your system (e.g. additional graphics capability) or they may appear to be copies of commercial software packages.

The unwitting user simply loads the program (the user is often asked to boot the system from the floppy disk on which the Trojan horse is distributed) and then some time later finds that things are not what they should be!

12.1.5 Stealth viruses

A stealth virus is designed to hide itself from any virus scanning protection that might be present. In an extreme case, the virus may be 'armoured' against the virus researcher trying to find out how it works and replicates itself.

12.1.6 Polymorphic viruses

A polymorphic virus is one which mutates each time it successfully invades a system. The changing nature of this type of virus makes them much harder to detect.

> **TIP:** Beware the unsolicited disk that arrives in your mail. There has been at least one major case of a Trojan horse distributed by an unscrupulous software company. Once installed this extremely unfriendly 'evaluation program' disables your system and then invites you to pay for the privilege of purchasing an antidote. Until you did this, your system is unusable. The (very) small print that accompanied the disk did, in fact, warn users that they would be committing themselves to a very large cash outlay when they installed the software. This is one reason why I immediately trash any unsolicited disks that arrive in my mail and I strongly advise that you do the same!

12.2 Email viruses

File attachments are one of the most common ways to get a virus infection on your machine. The damage is often to automatically email something nasty to all those in your address book. Do *not* open attachments that are executable, i.e. end with .vbs or .exe.

12.3 Hoaxes

Hoax emails can cause much alarm and hence damage. The *first* thing to do when you get a warning about a virus is this:

- Don't panic.
- Using the mouse, 'paint' a key part of the message like the 'gift from Microsoft' part of example 1 or 'jdbgmgr.exe' from example 2 (see below).
- Use CTRL C to copy the text.
- Start your web browser.
- Start the Google search engine by entering the URL, www.google.com
- Use CTRL V to paste the text into Google's search box. If there are spaces in the text, enclose all of it in quotes, e.g. 'gift from Microsoft'.
- See what you get. If the information you find is from Symantec, McAfee, etc. you can trust it. Most warnings are hoaxes and even experienced IT professionals can and do fall for them.

Also have a look at www.vmyths.com/

12.3.1 Example hoax virus warnings

Example 1: There is a deadly virus out there! It is called: 'Gift from Microsoft' or something like that! DO NOT OPEN!!! It will do major damage! I am telling you this from first hand account!

Example 2: The objective of this email is to warn all Hotmail users about a new virus that is spreading by MSN Messenger. The name of this virus is jdbgmgr.exe and it is sent automatically by the Messenger and by the address book too. The virus is not detected by McAfee or Norton and it stays quiet for 14 days before damaging the system.

The virus can be cleaned before it deletes the files from your system. In order to eliminate it, it is just necessary to do the following steps:

1. Go to Start, click 'Search'.
2. In the 'Files or Folders option' write the name jdbgmgr.exe
3. Be sure that you are searching in drive 'C'.
4. Click 'find now'.
5. If the virus is there (it has a little bear-like icon with the name of jdbgmgr.exe. DO NOT OPEN IT FOR ANY REASON.
6. Right click and delete it (it will go to the Recycle bin).
7. Go to the Recycle bin and delete it or empty the Recycle bin.

IF YOU FIND THE VIRUS IN ALL OF YOUR SYSTEMS SEND THIS MESSAGE TO ALL OF YOUR CONTACTS LOCATED IN YOUR ADDRESS BOOK BEFORE IT CAN CAUSE ANY DAMAGE.

TIP: If you find JDBGMGR.EXE on your computer, then it's probably not infected; but if you receive JDBGMGR.EXE as an email attachment, the attachment is probably infected!

12.4 Sources of viruses

Viruses can find their way into your system in a number of ways and from a variety of sources, including:

- Bulletin boards – Unless managed correctly, an open-access bulletin board can be an environment in which a virus can thrive. Most good

system operators arrange for regular or automatic virus checks. These can be instrumental in reducing the incidence of infection.
- The Internet and World Wide Web – When you download software from the World Wide Web (or receive an executable file along with an e-mail communication) you run the risk of transferring a virus to your computer.
- Networks – Like bulletin boards, networks must also be considered a 'public place' in which viruses can potentially thrive. Most network managers take steps to implement reliable anti-virus measures.
- Magazine cover disks – Even the magazine cover disk that proclaims that it has been 'tested for all known viruses' is not always what it seems. There have been several cases of viruses spread by cover disks even though the disk was 'checked' before duplication. So how did this happen? The important phrase is 'all known' – no scanning software can protect you from a virus strain that it doesn't know about!
- Pirated software – It should go without saying that pirate copies of commercial software passed from person to person carry a significant risk of being infected with viruses. Many viruses are spread this way.

TIP: It is illegal to make copies of commercial software and distribute these to others. People who indulge in this practice put themselves at risk not only of infecting their systems with viruses but also of prosecution for what amounts to theft. It just isn't worth it!

12.5 Virus prevention

Fortunately, there are plenty of things that you can do to safeguard yourself from virus infection and avoid the frustration of having to 'clean up' when disaster strikes. A few basic precautions will help you to avoid the vast majority of virus infections but if you do find that your system has been invaded by one of these 'nasties' you can usually destroy them quickly and easily using one of several proprietary 'virus killers'.

You can prevent viruses infecting your system by adopting 'clean habits', notably:

- Only download software from reputable bulletin board systems or shareware sites. If you are unsure about downloaded software you should scan it (using a proprietary anti-virus package) before you run any of the downloaded programs.
- Never use pirated copies of games and commercial software (these carry a very high risk of virus infection).
- Don't install or copy files onto your system from any unsolicited disk.

- Never install programs on your system from anything other than original distribution disks (or your own personally made backup copies of them).
- Make sure that other users of your system do not import their own software or install any software onto your system that you don't know about. Make them adhere to your 'clean habits'!
- Purchase a reputable anti-virus package. Use this to periodically scan your system and also to check any new disks that you are uncertain of.

TIP: You should ensure that your anti-virus software is regularly updated (monthly) to take account of any new virus strains that have appeared. Most anti-virus producers provide such a service at nominal cost.

TIP: In severe cases of virus infection you may find it necessary to reformat your hard disk and replace the programs and data stored on it in just the same way as you would if the hard disk itself had failed. This 'last resort' will eradicate an existing virus completely and should serve to further emphasize the need to make regular backups of the data stored on your hard disk. You can use one of several excellent proprietary backup packages to help you perform this task. Note, however, that if one (or more) of your backup files has been infected the problem will eventually recur when the file in question is executed.

12.6 Detecting and eliminating viruses

12.6.1 Detecting a virus

Some viruses will forcibly announce their presence by displaying messages, corrupting screen data, or simply by 'dissolving' your screen. Others are a little more subtle in operation. In such cases you should look for the following 'tell-tale' signs:

1. Has your hard disk increased in size for no apparent reason?
2. Have you noticed an increase (albeit small) in the size of individual programs?
3. Is hard disk activity occurring when you don't expect it?
4. Has your system become noticeably slower recently?
5. When your system accesses the floppy drive, does it take longer than before?

Next you should ask yourself what you have done recently that could have been responsible for importing a virus:

6. Have you installed any new software recently?
7. Have you recently downloaded software from the Web?
8. Have you transferred any programs from someone else's disks to your machine?
9. Has anyone else had access to your machine?
10. Has anyone recently sent you an email with an executable file attached?
11. Has anyone in your email address book either told you about odd emails from you (or even stopped talking to you!)

If you are sure that you are suffering a virus infection and not a hardware fault (you can confirm this using proprietary anti-virus software) the next stage is to eradicate the virus before it spreads.

12.6.2 Eliminating a virus

To prevent the virus present on your hard disk becoming active when you boot your system, you will need a 'clean boot' disk. This is simply an emergency boot disk which must, of course, have been produced before your system was infected.

You should boot from this disk (in which case your system will be 'clean' when the DOS prompt appears). You can then run a proprietary virus detection program to scan the complete system.

Scanning software will first check memory for any viruses that may have installed themselves into RAM. It will then check boot sectors, COM, EXE, overlay and data files for any viruses that may have attached themselves to your existing software.

This process may take some time as the scanning software works through the entire contents of your hard disk. When the scan has been completed, the anti-virus package will display a report on the types of virus (if any) that it has detected. If a virus is found, the scanning software may suggest the remedy. Alternatively, you may have to refer to a printed table in order to decide upon the best method for removing the virus. The technique will vary with the type of virus and, in particular, whether it invades boot sectors or attaches itself to files. In some cases (e.g. where the virus has overwritten some of your program code) you may have to reinstall applications software in order to restore the code to its original state.

12.7 Anti-virus software - seven of the best

A reliable anti-virus package is an essential purchase, particularly if you regularly exchange disks with other users or download software from the World Wide Web. There are currently many anti-virus software packages to choose from – the seven listed here are the current market leaders and any one of them will satisfy the needs of most users:

McAfee VirusScan	www.mcafee.com
Norton Antivirus	www.norton.com
PC-cillin	www.antivirus.com
Vexira Antivirus Persona	www.centralcommand.com
F-Secure Anti-Virus Personal Edition	www.f-secure.com
Norman Virus Control	www.norman.com
NOD32	www.nod32.com/home/home.htm

It is worth noting that several of the leading anti-virus companies provide virus scanning software that can be downloaded from the Internet for evaluation purposes. In many cases, this software is fully functional and it will allow you to tackle an immediate virus problem. Furthermore, when you become a registered user, you can also use the Web to periodically update your anti-virus software.

TIP: Not all anti-virus programs provide the same range facilities and it is important to check the features that you need before you commit to a particular product. As a minimum, you should consider on-access and on-demand scanning to be essential features. The ability to scan zipped files (i.e. files with a .ZIP extension) is highly desirable if you regularly download or exchange compressed files while the ability to detect macro viruses is essential if you regularly work with Microsoft Word and Excel files.

TIP: It is wise not to rely on cheap anti-virus programs as they may have unacceptably low recognition rates and may thus not detect all types of virus. It is also wise to ensure that your anti-virus software supplier will support the product with regular updates including the latest virus signatures. For this reason, it is wise to consider only purchasing well-known and well-supported programs from established suppliers.

TIP: When downloading software from the World Wide Web it is important to be aware that viruses may be present on any executable files (including those incorporated within zipped files) that you download. Viruses may also be present on executable macros (such as those available for Microsoft Word for Windows and Excel). When downloading *any* files from the Web, it is wise not to take any chances and ensure that each file is downloaded into a purpose created download directory and subsequently scanned before execution. Some, *but not all*, anti-virus packages are effective when used to scan zipped (.ZIP) files. If you do not have such a package, it is essential to unzip and then scan each file before the installation or setup procedure is started. You should also note that graphics files (e.g. .JPG and .GIF files) are non-executable and therefore they are not prone to virus infection.

12.8 Getting virus help via the Internet

In the event that you might need information and support in tracking down a virus, a connection to the World Wide Web can be invaluable. Some of the most useful sources of anti-virus information are listed below:

Computer Information Centre (CompInfo)
www.compinfo.co.uk
This UK-based website contains a large number of useful links including several to the sites of anti-virus software suppliers.

Dr Solomon's Software Virus Encyclopaedia
www.drsolomon.com/
This is another excellent on-line encyclopaedia of computer viruses which incorporates an alphabetic selection as well as a keyword search facility.

IBM Anti-Virus On-line
www.av.ibm.com/current/frontpage/
IBM's web pages include a number of useful articles dealing with anti-virus issues.

McAfee Virus Information (now the owners of Dr Solomon's)
www.drsolomon.com/
McAfee's website contains a great deal of information on a large number of known viruses.

Symantec Anti-Virus Research Centre
www.symantec.com/avcenter/vinfodb.html
Symantec's Anti-Virus Research Centre provides an excellent database of virus information.

13 Troubleshooting Windows error messages

There are fairly good hardware troubleshooters included with Windows 2000 and XP versions. Unfortunately, these are sometimes too general in nature so the section below sets out more detailed information.

Anyone who has been involved with PCs at anything more than the basic user level will almost certainly have come across the unhelpful (and occasionally totally incomprehensible) error messages that Windows, in all its incarnations, is capable of generating! This section is dedicated to helping you solve some of these problems by identifying the underlying causes of these frustrating and sometimes mind-bending problems. We also show you how to make use of the excellent, but little known, Dr Watson troubleshooting tool that Microsoft supplies as part of its Windows operating system.

The problems that can occur in Windows can be arranged into the following main categories:

- Invalid page faults.
- General protection faults.
- Fatal exceptions.
- Protection errors.
- Kernel errors.

Note that, while the information in this section applies generally to all versions of Windows 9x (i.e. Windows 95, Windows 98 and Windows 98SE) it should be applied with some caution to Windows ME and Windows 3.x.

At this stage it's worth noting that modern CPUs are designed to detect situations in which an executable program attempts to do something that is nonsensical or 'invalid' in terms of the hardware and software configuration of the system. The most common problems are stack faults, invalid instructions, divide errors (divide by zero) and general protection faults. These generally indicate non-standard code in a program. The following are examples of faults that can occur in a Windows-based program, in Windows itself, or in a Windows device driver (for example, a video adapter driver), as well as the implications for the CPU.

Stack Fault (Interrupt 12)

Reasons for a stack fault include:

- An instruction tries to access memory beyond the limits of the stack segment (POP, PUSH, ENTER, LEAVE, or a stack relative access: MOV AX, [BP+6]).
- Loading SS with a selector marked not present, but otherwise valid (this should not happen under Windows).

Note that stack faults are always fatal to the current program in Windows.

Invalid Instruction (Interrupt 6)

The CPU detects most invalid instructions, and generates an interrupt. This is always fatal to the program. This should never happen, and is usually caused by running data instead of code.

Divide Error (Interrupt 0)

This occurs when the destination register cannot hold the result of a divide operation. This could be caused by an attempt to divide by zero, or a divide overflow.

General Protection Fault (Interrupt 13)

All protection violations that do not cause another exception cause a general protection exception. This includes, but is not limited to:

- Exceeding the segment limit when using the CS, DS, ES, FS, or GS segments. This is a very common problem in programs and it is usually caused when a program miscalculates how much memory is required in an allocation.
- Transferring execution to a segment that is not executable (for example, jumping to a location that contains garbage).
- Writing to a read-only or a code segment.
- Loading a bad value into a segment register.
- Using a null pointer. A value of zero (i.e. 0) is defined as a null pointer. When operating in protected mode, it is always invalid to use a segment register that contains zero.

TIP: A great deal of hardware related information is available from the Microsoft Support website. You can view this information and search the articles available by visiting the website at:

support.microsoft.com/support/windows/topics/hardware/hwddresctr.asp

13.1 Invalid page faults

This error message can occur for any of the following reasons:

- An unexpected event has occurred in Windows. An invalid page fault error message often indicates that a program improperly attempted to use random access memory (RAM). For example, this error message can occur if a program or a Windows component reads or writes to a memory location that is not allocated to it. When this behaviour occurs, the program can potentially overwrite and corrupt other program code in that area of memory.
- A program has requested data that is not currently in virtual memory, and Windows attempts to retrieve the data from a storage device and load it into RAM. An invalid page fault error message can occur when Windows cannot locate the data. This behaviour often occurs when the virtual memory area becomes corrupted.
- The virtual memory system has become unstable because of a shortage of physical memory (RAM).
- The virtual memory system has become unstable because of a shortage of free disk space.
- The virtual memory area has been corrupted by a program.
- A program is attempting to access data that is being modified by another program that is running.

If you are using Windows 95 or Windows 98, you may receive the following error message:

> This program has performed an illegal operation and will be shut down. If the problem persists, contact the progrtam vendor.

If you subsequently click on Details, you may receive an error message of the form:

> [Program] caused an invalid page fault in module at [location].

This type of error is 'unrecoverable' and hence, after you click OK, the program somewhat unhelpfully shuts down!

Note that if you are using Windows ME (Millennium Edition), you will receive an error message of the form:

> [Program] has caused an error in [address]. [Program] will now close.

If you continue experiencing this type of error message you should restart the computer. To view the details of the problem you should press ALT + D, or open the Faultlog.txt file in the Windows folder.

To resolve this problem it is important to identify when, and in what situation, the error message *first* occurred. Also, determine if you recently made changes to the computer, for example if you installed software or changed the hardware configuration. In either case, you should use a clean boot troubleshooting procedure (see later) to help you identify the cause of the error message.

> **TIP:** For additional information about invalid page fault error messages, view the following Microsoft Knowledge Base article: Q286180 *Invalid Page Fault Errors occur in Windows.*

13.2 General protection faults

A general protection fault (GP fault) often indicates that there is a problem with the software that you are using or that you need to update a device driver installed on your computer. The Dr Watson tool can often help you to identify the cause of the error message by taking a snapshot of your computer when the fault occurs. So, if you encounter a GP fault, you should run the Dr Watson tool so that you can 'catch' the error the next time that it occurs (see later).

Because general protection faults can be caused by software or hardware, the first step is to restart your computer in Safe mode in order to narrow down the source of the error. Restarting in Safe mode will allow you to check whether the problem is due to either:

- Hardware or Windows core files, or
- A driver or application program.

Restarting in Safe mode allows you to test your computer in a state in which only essential components of Windows are loaded. If you restart your computer in Safe mode and the error message does not occur, the origin is more likely to be a driver or program. If you restart in Safe mode and then test your computer and the error message does occur, the issue is more likely to be hardware or damaged Windows core files.

13.2.1 Testing in Safe mode

If you are not using Dr Watson (see later) the following procedure is recommended:

1. Enter Safe mode, as follows:
   (a) for Windows 95, restart your computer, press F8 when you see the 'Starting Windows 95' message, and then choose Safe Mode
   (b) for Windows 98, restart your computer, press and hold down the CTRL key until you see the Windows 98 Startup menu, and then choose Safe Mode
   (c) for Windows Millennium Edition (ME), press and hold down the CTRL key while you restart the computer, and then choose Safe Mode on the Windows ME Startup menu
2. Test your computer in Safe mode. If the error does not occur, use the appropriate steps below for your operating system. If the error does occur, it is likely to be caused by a problem with your Windows installation or you may be experiencing a symptom of faulty hardware.
3. After your computer restarts in Safe mode, use the System Configuration Utility tool (Msconfig.exe) to minimize conflicts that may be causing the problem:
   (a) click Start, point to Programs, point to Accessories, point to System Tools, and then click System Information
   (b) on the Tools menu, click System Configuration Utility
   (c) on the General tab, click Selective Startup, and then click to clear the following check boxes:
      - Process Config.sys File
      - Process Autoexec.bat File
      - Process Winstart.bat File (if available)
      - Process System.ini File
      - Process Win.ini File
      - Load Startup Group Items
4. Click OK, and then restart your computer normally when you are prompted. After you restart and test your computer, if you still do not receive the error message, continue with the next steps:
   (a) run the System Configuration Utility tool, click to select one item in the Selective Startup box, click OK, and then restart your computer and test
   (b) continue selecting items using the System Configuration Utility tool until all of the items in the Selective Startup box are selected. If you select an item and your issue recurs, click the tab of the corresponding item in Selective Startup, clear

half of the check boxes, click OK, and then restart your computer

(c) continue this process until you narrow down the setting that is causing your problem. If you can restart your computer successfully when all items are checked, run the System Configuration Utility tool, click to select Normal Startup, click OK, and then restart your computer

> **TIP:** For additional information about using Msconfig.exe, view the following Microsoft Knowledge Base article: Q192926 *How to Perform Clean-Boot Troubleshooting for Windows 98.*

13.2.2 Real mode configuration problems

The following steps can help you to determine if the problem that you are experiencing is due to the real mode configuration of your computer. This could include drivers that are loaded from your Config.sys and Autoexec.bat files.

1. Restart your computer. When the 'Starting Windows' message is displayed, press F8, and then click Step-By-Step Confirmation from the Startup menu.
2. When you are prompted, load the following items (if you are prompted to load any other items, press N):
   (a) Dblspace driver
   (b) Himem.sys
   (c) Ifshlp.sys
   (d) Dblbuff.sys
3. Load the Windows graphic user interface (GUI), choosing to load all Windows drivers.

> **TIP:** Windows 95 does not require the Config.sys and Autoexec.bat files, but some tools installed on the computer may require them. You should never rename the Config.sys and Autoexec.bat files until you perform a successful interactive boot to verify that they are not needed.

13.2.3 Startup conflicts

If the clean boot of your real mode configuration eliminates the problem, isolate the conflict with a terminate-and-stay-resident (TSR) or real mode device driver by using the Step-By-Step Confirmation function.

Load Windows by booting to a command prompt and starting Windows by typing win, holding down the SHIFT key for the duration of the boot. This prevents any programs from loading automatically at startup.

If the issue is resolved by bypassing the Startup group, remove each of the programs from the Startup group individually to isolate the program that is causing the problem. Note that you can prevent programs from loading by removing the program's string from the following registry keys:

- HKEY_LOCAL_MACHINE\SOFTWARE\Microsoft\Windows\CurrentVersion\Run
- HKEY_LOCAL_MACHINE\SOFTWARE\Microsoft\Windows\CurrentVersion\RunServices

Programs may also be loading from the following registry key:

- HKEY_CURRENT_USER\Software\Microsoft\Windows\CurrentVersion\Run

Note that, within the Win.ini file, the 'load = ' and 'run = ' lines can also be used to start programs automatically.

> **TIP:** If you use Registry Editor incorrectly, you may cause serious problems that may require you to reinstall your operating system. It is therefore essential that you use the Registry Editor with care, ensuring that you have a backup of the registry data file that you can use to reinstate the registry in the event that a problem does occur!

13.2.4 Faulty Windows configuration files

To test the Windows configuration files, you should use the following steps:

1. Boot to a command prompt.
2. Rename the Win.ini file by typing the following DOS command:

   ren c:\windows\win.ini *.bak

3. Start Windows by typing win.
4. If this procedure corrects the problem, ensure that the 'load = ' and 'run = ' lines in the Windows section of the Win.ini file are either blank or preceded with a remark semicolon (;) in order to prevent the items from loading.
5. Rename the System.ini file by typing the following command:

   ren c:\windows\system.ini *.bak

6. Windows 95 requires a System.ini file to load the graphic user interface. Replace the original file by typing the following DOS command:

   copy c:\windows\system.cb c:\windows\system.ini

7. Windows 95 does not load a mouse driver with the System.cb file. In order to re-establish contact with the mouse you will need to edit the new System.ini file by adding the following lines:

   [386Enh]
   mouse = *vmouse, msmouse.vxd

   [boot]
   drivers = mmsystem.dll
   mouse.drv = mouse.drv

8. Next start Windows by typing win at the command prompt. If replacing the original System.ini file with the System.cb file corrects the issue, the problem most likely resides with either the [boot] or [386Enh] sections of the original System.ini file.
9. Finally, restore the original file in order to troubleshoot it. To isolate the cause of the problem, place a remark semicolon (;) at the beginning of each line in turn to prevent the item from loading. This will allow you to identify the faulty item.

> **TIP:** For additional information about the System.ini file and its default entries, view the article in the Microsoft Knowledge Base: Q140441 *Creating a New System.ini File Without Third-Party Drivers*

> **TIP:** The Winstart.bat file is used to load TSRs that are required for Windows-based programs and are not needed in MS-DOS sessions.

13.2.5 Protected mode device drivers

It is important to note that Safe mode disables all protected mode device drivers for Windows 95. You can conduct testing for incompatible components and resource conflicts by disabling the protected mode device drivers in Device Manager. The following procedure can be used to remove protected mode device drivers:

1. Click Start, point to Settings, click Control Panel, and then double-click System.
2. On the Device Manager tab, click View Devices By Type.

3. Disable each of the protected mode device drivers. For example:
   (a) double-click the Floppy Disk Controllers branch to expand it
   (b) click Standard Floppy Disk Controller, and then click Properties
   (c) on the General tab, click to clear the Original Configuration (Current) check box, and then click OK

Note that if you have enabled hardware profiles, there is a check box for each of the configurations. Clear the check box for the hardware profile you are troubleshooting.

4. Repeat steps 1 to 3 for each device in Device Manager. Click Close, and then restart the computer.

If you resolve the issue by disabling the protected mode drivers in Device Manager, you may have a hardware conflict or a driver may be incompatible with your hardware. If you determine that a protected mode device driver is incompatible with your hardware you will need to contact the hardware manufacturer in order to determine the availability of a new driver.

13.2.6 Video adapter problems

The following steps are required to change the video driver:

1. In order to return to your original video settings you should first do the following:
   (a) back up the System.ini file
   (b) note the current desktop area (resolution) and colour palette
   (c) note the name of your current video adapter

2. To change to the VGA video driver, follow these steps:
   (a) start Windows in Safe mode
   (b) click Start, point to Settings, click Control Panel, and then double-click Display
   (c) on the Settings tab, click Change Display Type
   (d) in the Adapter Type area, click Change
   (e) click Show All Devices
   (f) in the Manufacturers box, click (Standard Display Types)
   (g) in the Models box, click Standard Display Adapter (VGA), and then click OK
   (h) click OK or Close until you return to Control Panel
   (i) restart the computer

If you resolve the issue and determine that your video driver is incompatible with your system you will need to contact the video adapter manufacturer to determine the availability of a new driver.

> **TIP:** The easiest way to locate the correct driver (or to update a driver to the latest version) is to download it from the hardware manufacturer's website.

> **TIP:** Safe mode starts Windows with a basic VGA video driver. To determine if the issue you are experiencing is related to your video driver, change to the VGA driver for testing purposes. Note, however, that if you have removed the protected mode drivers in order to isolate conflicts (as described previously) you will have already reverted back to the basic VGA video driver.

13.2.7 Registry problems

When you start Windows in Safe mode the registry is only partially read. Damage to the registry may not therefore be evident when running in Safe mode and you may need to replace the existing registry data file (System.dat) with a recent backup in order to see if this resolves the problem in which case the cause is likely to be a damaged registry data file. The following procedure is required in order to troubleshoot a damaged registry:

1. Boot to a command prompt.
2. Remove the file attributes from the backup of the registry by typing the following DOS command:

   ```
   c:\windows\command\attrib -h -s -r c:\system.1st
   ```

3. Remove the file attributes from the current registry by typing the following DOS command:

   ```
   c:\windows\command\attrib -h -s -r c:\windows\system.dat
   ```

4. Rename the registry by typing the following command:

   ```
   ren c:\windows\system.dat *.dax
   ```

5. Copy the backup file to the current registry by typing the following command:

   ```
   copy c:\system.1st c:\windows\system.dat
   ```

6. Restart the computer.

TIP: The System.1st file is a backup of the registry that was created during the final stage of the original Windows Setup. Therefore, the 'Running Windows for the first time' banner is displayed and Windows will finalize its settings as if it is being installed for the first time.

NOTE: If replacing the System.dat file with the System.1st file resolves the issue, the problem may be related to registry damage. Any programs and device drivers that were subsequently installed may require reinstallation to update the new registry. For this reason it is essential to keep all of your original installation disks in a safe place!

If you determine that the problem is not caused by a faulty registry data file you will need to restore the original registry data file. The procedure is as follows:

1. Restart the computer to a command prompt.
2. Type the following commands, pressing ENTER after each command:

   ```
   c:\windows\command\attrib -s -h -r c:\windows\system.dat
   copy c:\windows\system.dax c:\windows\system.dat
   ```

3. Overwrite the existing System.dat file if you are prompted to do so.
4. Restart the computer.
5. If the problem is still unresolved, the next stage is that of reinstalling the Windows core files. You will need the original installation CD-ROM and you should install Windows in a 'clean' folder. If the new installation resolves the problem this usually indicates that either one or more of your Windows core files has been damaged, or that there is an error in the configuration of your original installation. You can choose to use the new installation of Windows, but you will have to reinstall any application programs so that they are correctly recognized by Windows.
6. If the problem is not resolved with a 'clean' installation, the condition is probably attributable to faulty hardware. In such a case you may need to contact the motherboard manufacturer as well as the manufacturer of any adapter cards that are fitted to the system. If you have access to a similar system that is fault-free, you should, of course, be able to carry out substitution tests.

13.3 Fatal exceptions

Fatal exceptions occur in the following situations:

- If access to an illegal instruction has been encountered.
- If invalid data or code has been accessed.
- If the privilege level of an operation is invalid.

When any of these situations occur, the processor returns an exception to the operating system, which in turn is handled as a fatal exception error message. In many situations, the exception is non-recoverable and you must either shut down or restart the computer, depending on the severity of the error.

Fatal exceptions are likely to be encountered when:

- You attempt to shut down the computer.
- You start Windows.
- You start an application or other program from within Windows.

In either of these cases, an error message like that shown below will appear:

```
A fatal exception [code] has occurred at [location].
```

In order to distinguish the type of fatal exception that has occurred these errors are given codes that are returned by a program. The value of the code represents the enhanced instruction pointer to the code segment; the 32-bit address is the actual address where the exception occurred.

It is important to appreciate that, while Windows does not actually cause these errors, it has the exception-handling routine for that particular processor exception and this, in turn, is what actually displays the error message.

For those with some experience of low-level architecture, the various fatal exception error codes (in hexadecimal) are listed below:

00: Divide Fault

The processor returns this exception when it encounters a divide fault. A divide fault occurs if division by zero is attempted or if the result of the operation does not fit in the destination operand.

02: NMI Interrupt

Interrupt 2 is reserved for the hardware non-maskable-interrupt condition. No exceptions trap via interrupt 2.

04: Overflow Trap

The overflow trap occurs after an INTO instruction has executed and the 0F bit is set to 1.

05: Bounds Check Fault

The BOUND instruction compares the array index with an upper and lower bound. If the index is out of range, then the processor traps to interrupt 05.

06: Invalid Opcode Fault

This error is returned if any one of the following conditions exists:

- The processor tries to decode a bit pattern that does not correspond to any legal computer instruction.
- The processor attempts to execute an instruction that contains invalid operands.
- The processor attempts to execute a protected mode instruction while running in virtual 8086 mode.
- The processor tries to execute a LOCK prefix with an instruction that cannot be locked.

07: Coprocessor Not Available Fault

This error occurs if the computer does not have a math coprocessor and the EM bit of register CR0 is set indicating that numeric data processor emulation is being used. Each time a floating point operation is executed, an interrupt 07 occurs.

This error also occurs when a math coprocessor is used and a task switch is executed. Interrupt 07 tells the processor that the current state of the coprocessor needs to be saved so that it can be used by another task.

08: Double Fault

Processing an exception sometimes triggers a second exception. In the event that this occurs, the processor will issue an interrupt 08 for a double fault.

09: Coprocessor Segment Overrun

This error occurs when a floating point instruction causes a memory access that runs beyond the end of the segment. If the starting address of the floating point operand is outside the segment, then a general protection fault occurs (interrupt 0D).

10 (0Ah): Invalid Task State Segment Fault

Because the task state segment contains a number of descriptors, any number of conditions can cause exception 0A. Typically, the processor can gather enough information from the task state segment to issue another fault pointing to the actual problem.

11 (0Bh): Not Present Fault

The not present interrupt allows the operating system to implement virtual memory through the segmentation mechanism. When a segment is marked as 'not present', the segment is swapped out to disk. The interrupt 0B fault is triggered when an application needs access to the segment.

12 (0Ch): Stack Fault

Stack fault occurs with error code 0 if an instruction refers to memory beyond the limit of the stack segment. If the operating system supports expand-down segments, increasing the size of the stack should alleviate this problem. Loading the stack segment with invalid descriptors will result in a general protection fault.

13 (0Dh): General Protection Fault

Any condition that is not covered by any of the other processor exceptions will result in a general protection fault. The exception indicates that this program has been corrupted in memory, usually resulting in immediate termination of the program.

14 (0Eh): Page Fault

The page fault interrupt allows the operating system to implement virtual memory on a demand-paged basis. An interrupt 14 is usually issued when an access to a page directory entry or page table with the present bit set to 0 (not present) occurs. The operating system makes the page present (usually retrieves the page from virtual memory) and reissues the faulting instruction, which then can access the segment. A page fault also occurs when a paging protection rule is violated (when the retrieve fails, or data retrieved is invalid, or the code that issued the fault broke the protection rule for the processor). In these cases the operating system takes over for the appropriate action.

16 (10h): Coprocessor Error Fault

This interrupt occurs when an unmasked floating-point exception has signalled a previous instruction. (Because the 80386 does not have access to the floating-point unit, it checks the ERROR pin to test for this condition.) This is also triggered by a WAIT instruction if the emulate math coprocessor bit at CR0 is set.

17 (11h): Alignment Check Fault

This interrupt is only used on the 80486 CPUs. An interrupt 17 is issued when code executing at ring privilege 3 attempts to access a word operand that is not on an even-address boundary, a double-word operand that is not divisible by four, or a long real or temp real whose address is not divisible by eight. Alignment checking is disabled when the CPU is first powered up and is only enabled in protected mode.

Because there are many conditions that can cause a fatal exception error, the first step in resolving the issue is to narrow the focus by using the clean boot procedure described earlier. It is also worth noting that many problems occur because of conflicting drivers, terminate-and-stay-resident programs (TSRs), and other settings that are loaded when the computer first starts.

13.4 Protection errors

Windows Protection error messages occur when a computer attempts to load or unload a virtual device driver (VxD). This error message is a way to let you know that there is a problem with the device driver. In many cases, the VxD that did not load or unload is mentioned in the error message. In other cases, you may not be able to determine the VxD that caused the behaviour; however, you should be able to find the cause of the error message if you use clean boot troubleshooting.

Windows Protection error messages can occur in any of the following situations:

- If a real mode driver and a protected mode driver are in conflict.
- If the registry is damaged.
- If either or both the Win.com file or the Command.com file are infected with a virus, or if either of the files has become corrupted or damaged.
- If a protected mode driver is loaded from the System.ini file and the driver is already initialized.
- If there is a physical input/output (I/O) address conflict or a random access memory (RAM) address conflict.

- If there are incorrect complementary metal oxide semiconductor (CMOS) settings for a built-in peripheral device (such as cache settings, CPU timing, hard disks, and so on).
- If the plug and play feature of the basic input/output system (BIOS) on the computer is not working correctly.
- If the computer contains a malfunctioning cache or malfunctioning memory.
- If the motherboard on the computer is not working properly.

When you start Windows, you may receive one of the following error messages:

```
While initializing device [device name] Windows Protection Error
```

or the even more succinct (and somewhat less helpful) message:

```
Windows Protection Error
```

When you shut down the computer, you may receive the following error message:

```
Windows Protection Error
```

The following procedure is recommended when investigating Windows Protection errors:

1. First enter Safe mode, as follows:
   (a) for Windows 95, restart your computer, press F8 when you see the 'Starting Windows 95' message, and then choose Safe Mode
   (b) for Windows 98 (and Windows 98 Second Edition), restart the computer, press and hold down the CTRL key until you see the Windows 98 Startup menu, and then choose Safe Mode
   (c) for Windows Millennium Edition (ME), press and hold down the CTRL key while you restart the computer, and then choose Safe Mode on the Windows ME Startup menu
2. If you do not receive the error message when you start the computer in Safe mode (or when you shut down the computer from Safe mode) you should follow the procedure described earlier in order to check that the computer is correctly configured and that the system hardware and associated drivers are operating correctly.
3. If you receive the error message when you attempt to start the computer in Safe mode, you should follow the steps listed below to restore the registry:

(a) boot to a command prompt
(b) remove the file attributes from the backup of the registry by typing the following DOS command:

c:\windows\command\attrib -h -s -r c:\system.1st

(c) remove the file attributes from the current registry by typing the following DOS command:

c:\windows\command\attrib -h -s -r c:\windows\system.dat

(d) rename the registry by typing the following command:

ren c:\windows\system.dat *.dax

(e) copy the backup file to the current registry by typing the following command:

copy c:\system.1st c:\windows\system.dat

4. Restart the computer and verify that the computer's current CMOS settings are correct.
5. Install a 'clean' copy of Windows in an empty folder. If the new installation resolves the problem this usually indicates that either one or more of your Windows core files has been damaged, or that there is an error in the configuration of your original installation. You can choose to use the new installation of Windows, but you will have to reinstall any application programs so that they are correctly recognized by Windows.
6. If the problem is not resolved with a 'clean' installation, the condition is probably attributable to faulty hardware. In such a case you may need to contact the motherboard manufacturer as well as the manufacturer of any adapter cards that are fitted to the system. If you have access to a similar system that is fault-free, you should, of course, be able to carry out substitution tests.

TIP: The virtual device driver (VxD) that is generating the error message can be any VxD, either a default VxD that is installed, or a third-party .386 driver that is loaded from the System.ini file. If you do not know which driver is causing the error message, create a Bootlog.txt file, and then check to see which driver is the last driver that is initialized. This is typically the driver that is causing the problem.

TIP: You may also receive a Windows Protection error message when you restart Windows after you install a program or make a configuration change to your computer. For additional information about this problem, view the following article in the Microsoft Knowledge Base: Q157924 *Err Msg: 'IOS Failed to Initialize' on Boot.*

13.5 Kernel errors

The Kernel32.dll file is a 32-bit dynamic link library file that is found in Windows 95, Windows 98 and Windows Millennium Edition (ME). The Kernel32.dll file handles memory management, input/output operations and interrupts. When you start Windows, Kernel32.dll is loaded into a protected memory space so that other programs do not take over that memory space.

On occasion, you may receive an invalid page fault (IPF) error message. This error message occurs when a program tries to access the Kernel32.dll protected memory space. Occasionally, the error message is caused by one particular program while on other occasions it may be generated by several programs.

If the problem results from running one program, the program needs to be replaced. If the problem occurs when you access multiple files and programs, the damage is likely caused by damaged hardware. You may want to clean boot the computer to help you identify the particular third-party memory resident software. Note that programs that are not memory resident can also cause IPF error messages.

The following faults can cause Kernel32.dll error messages:

- Damaged swap file.
- File allocation damage.
- Damaged password list.
- Damaged or incorrect version of the Kernel32.dll file.
- Damaged registry.
- Hardware, hot CPU, overclocking, faulty broken power supply, RF noise, or a defective hard disk controller.
- BIOS settings for wait states, RAM timing, or other BIOS settings.
- Third-party software that is damaged or incorrectly installed .dll files that are saved to the desktop.
- A non-existent or damaged Temp folder.
- A corrupted control panel (.cpl) file.
- Incorrect or damaged hardware driver.
- Incorrectly installed printer drivers (or HP Jetadmin drivers).
- Damaged Java machine.
- Damaged .log files.
- Damaged entries in the History folder.
- Incompatible or damaged dynamic link library files.
- Viruses.

- Damaged or incorrect Msinfo32.exe file.
- Low disk space.

If you are using Windows 95 or Windows 98, you may receive the following error message:

> This program has performed an illegal operation and will be shut down. If the problem persists, contact the program vendor.

When you click Details, you may receive the following error message:

> [Program] caused an invalid page fault in module at [location]

After you click OK, the program shuts down.

If you are using Windows Millennium Edition (ME), you may receive the following error message:

> [Program] has caused an error in [location].
> [Program] will now close.

To view the details, press ALT + D, or open the Faultlog.txt file in the Windows folder. If you continue experiencing problems, you should try restarting your computer.

13.6 Dynamic link library faults

A dynamic link library (DLL) file is an executable file that allows programs to share code and other resources necessary to perform particular tasks. Microsoft Windows provides DLL files that contain functions and resources that allow Windows-based programs to operate in the Windows environment.

DLLs usually have a .DLL extension; however, they may also have an .EXE or other extension. For example, Shell.dll provides the object linking and embedding (OLE) drag and drop routines that Windows and other programs use while Kernel.exe, User.exe and Gdi.exe are examples of DLLs with .EXE extensions and they all provide code, data or routines to programs running under the Windows operating system. In Windows, an installable driver is also a DLL. A program can open, enable, query, disable and close the driver based on instructions written in the DLL file.

DLLs may be found in the Windows directory, Windows\System directory or in a program's directory. If a program is started and one of its DLL files is missing or damaged, you may receive an error message like:

```
Cannot find [filename.dll]
```

If a program is started with an outdated DLL file or mismatched DLL files, the error message

```
Call to undefined dynalink
```

may be displayed. In these situations, the DLL file must be obtained and placed in the proper directory in order for the program to run correctly.

The following procedure can be used to determine the version number, company name or other information about a dynamic link library file:

1. Click Start, point to Find, and then click Files or Folders.
2. In the Name box, type the name of the file you want to find, for example, 'shell32.dll' (but without the quotation marks).
3. Click Local Hard Drives (or the drive letter you want to search) in the Look In box, and then click Find Now.
4. Right-click the file in the list of found files, click Properties, and then click the Version tab.

13.7 Using Dr Watson

The diagnostic tool, Dr Watson, is supplied as part of the Windows operating system yet rarely is it ever referred to and most Windows users don't know that it exists! If a program fault occurs, Dr Watson will generate a snapshot of the current software environment which can provide invaluable information of what was happening at the point at which the fault occurred.

To start Dr Watson, you can either:

1. Click Start, click Run.
2. Enter 'drwatson' (without the quotation marks) in the box and then click on OK.

or

1. Click Start, select Programs and Accessories, and then click on System Tools.
2. Click System Information, and then click Dr Watson on the Tools.

When Dr Watson is running in the background you will see an additional icon displayed on your taskbar.

You can click the Details button in the error message to view the information that is gathered by Dr Watson. However, in most cases you will want to have a record of what was happening at the point at which the fault occurred. If this is the case, you can generate a log file by double-clicking the Dr Watson icon on the taskbar. In either case, Dr Watson gathers information about the operating system and then a Dr Watson dialogue box is displayed.

The log files produced by Dr Watson have a .wlg extension and they are stored in the \Windows\Drwatson folder. The log file provides a great deal of useful information including the name of the program that has created the fault, the program that the fault occurred in (not necessarily the same), and the memory address where the fault occurred. It is important to note that Dr Watson cannot create a snapshot if the program does not respond (i.e. if it hangs).

When you run Dr Watson (Drwatson.exe), it collects detailed information about the state of your operating system at the time of a program fault. Dr Watson then intercepts the software faults, identifies the software that has produced the fault, and then provides a detailed description of the cause. When this feature is enabled, Dr Watson automatically logs this information.

When Dr Watson is loaded, click any tab to move out of the text box. The Dr Watson window closes if you press ENTER. To view the advanced tabs in Dr Watson, follow these steps:

1. Double-click the Dr Watson icon.
2. On the View menu, click Advanced View.

The following tabs will then be displayed (see Figure 13.1) providing detailed information about the system:

System — Includes information that you would see on the General tab of System Properties.

Tasks — Includes information about the tasks that were running when the snapshot was taken. This tab also includes information about the program, the version, the manufacturer, the description, the path, the type and the program that this program is related to (when this information is available). (See Figure 13.2.)

Startup — Includes information about the programs that are configured to load during Startup. This tab includes the program name, and information about where the program was loaded from, and the command line that is used to load the program. (See Figure 13.3.)

Hooks — Provides information about modules that have intercepted (i.e. 'hooked') various aspects of the system. This tab can be used to show the hook type, the application and the path. (See Figure 13.4.)

Kernel Drivers Includes information about where the Kernel mode drivers are installed, including the name of the driver, the version, the manufacturer, the description, the likely path, information about where the driver is loaded from, the type of driver and the program that the driver related to (when information is available). (See Figure 13.5.)

User Drivers Includes information about the User mode drivers that are installed, including the name of the driver, the version, the manufacturer, the description, the likely path, the type of driver and the program that the driver is related to (when information is available). (See Figure 13.6.)

MS-DOS Drivers Includes information about the MS-DOS drivers that are installed. (See Figure 13.7.)

16-bit Modules Includes information about the 16-bit modules that were in memory when the snapshot was taken, including the name of the module, the version, the manufacturer, the description, the likely path, the type of driver and the program that the driver is related to (when information is available). (See Figure 13.8.)

Details Lists the events that occurred before and during the fault, in progressive order. Note that this tab is only displayed when Dr Watson has captured a fault.

If you experience a program fault, and you want to use Dr Watson, follow these steps:

1. Try to reproduce the fault to verify that it is not a random failure.
2. Click Start, point to Programs, point to Accessories and then click System Tools.
3. Click System Information, and then on the Tools menu, click Dr Watson.
4. Reproduce the fault.
5. Click Details in the Program Fault window.
6. View the Diagnosis window to determine the source of the fault.
7. If the issue is intermittent or not easy to reproduce, put Dr Watson in your Startup folder so that it is always running and will be ready to capture the fault information as and when the fault recurs.
8. When the fault next occurs examine the information captured in the log file. To save the information generated by Dr Watson, click Save on the File menu. You may also wish to add a few comments of your own stating under what circumstances the fault

occurred. When you have done this, select the File menu and click Save or Save As to save the file. Note that if you only click OK in the Dr Watson dialogue box, the information that you enter in the text box is not saved.

9. You can later view a Dr Watson log file by using the Dr Watson program or by using Microsoft System Information (MSInfo). To view Dr Watson log files by using MSInfo, follow these steps:
   (a) click Start, point to Programs, point to Accessories, point to System Tools and then click System Information
   (b) on the File menu, click Open
   (c) open the folder where the Dr Watson log is saved
   (d) in the Files of type list, click Dr Watson Log File (*.wlg)
   (e) click the file, and then click Open
10. To print Dr Watson log files, click on Print from the File menu. To print only specific information, you can use Microsoft System Information to view the log file, and then copy the specific information to an ASCII text editor, such as Microsoft Notepad. (Note that, depending on the software that happens to be running, a typical Dr Watson log file can amount to more than 15 pages of A4 text!)

Dr Watson can be configured using the limited number of options available (see Figure 13.9). The procedure for customizing Dr Watson to your own requirements is as follows:

1. Select the View menu and click Options.
2. Click on Log Files to configure the number of log files that are able to be stored on the computer and the folder that the log files will be saved in.
3. Click on Disassembly to configure the number of CPU instructions and stack frames that are to be reported in the log file.
4. Click on View to configure the view that Dr Watson is displayed in (either Standard View or Advanced View).

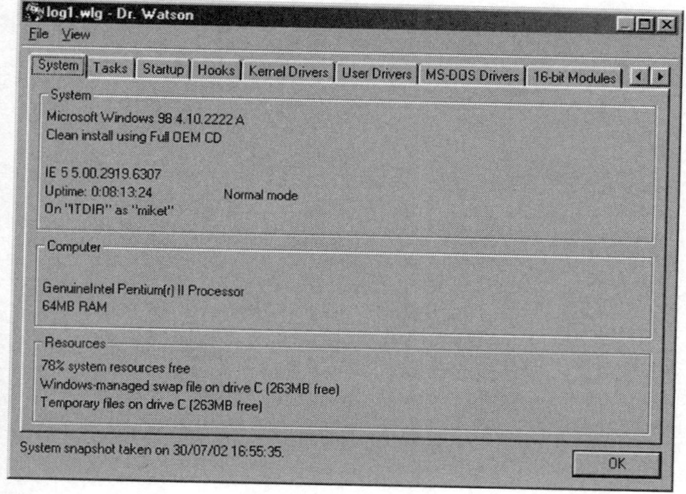

Figure 13.1 The Dr Watson dialogue box. Dr Watson has captured system information in a file named log1.wlg (see top left of window). The System tab displays version data relating to Windows and its installation (in this case, a clean installation using a full OEM CD), the version of Internet Explorer, the current log-in information (user name), the hardware platform (Pentium II processor with 64 MB RAM), and the available resources (78% free, 263MB free space on the C: drive, etc.)

TIP: You can configure Dr Watson to load automatically when Windows starts. To do this, create a shortcut to Drwatson.exe in the Startup folder. This configuration is useful when an issue is not easily reproducible. When Dr Watson traps the program fault and creates the log, you can contact technical support for further assistance.

TIP: Dr Watson is best used with reproducible faults. With intermittent faults you may often not be able to determine the cause of the fault, in which case you should follow the procedures described earlier depending on the exact nature of the Windows error message that has been generated.

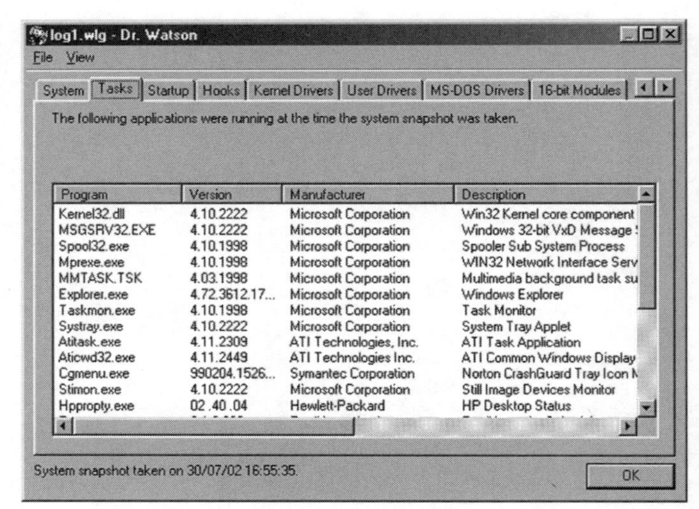

Figure 13.2 The Tasks tab displays a list of the programs that were running at the point at which the snapshot was taken. This important information shows the filename of the executable as well as its version number, its manufacturer and a brief description that tells you what it does

> **TIP:** The Dr Watson dialogue box includes a text box that you can use to enter information about what was happening when the fault occurred. This information can be extremely useful later particularly when the same machine next produces errors.

> **TIP:** When a program fault occurs, the Dr Watson log file is automatically named Watson.wlg (where is an incremented number). By default, Dr Watson log files are saved to the \Windows\Drwatson folder.

> **TIP:** For additional information about using Dr Watson, view the following Microsoft Knowledge Base article: Q275481 *How to Troubleshoot Program Faults with Dr Watson.* www.support.microsoft.com

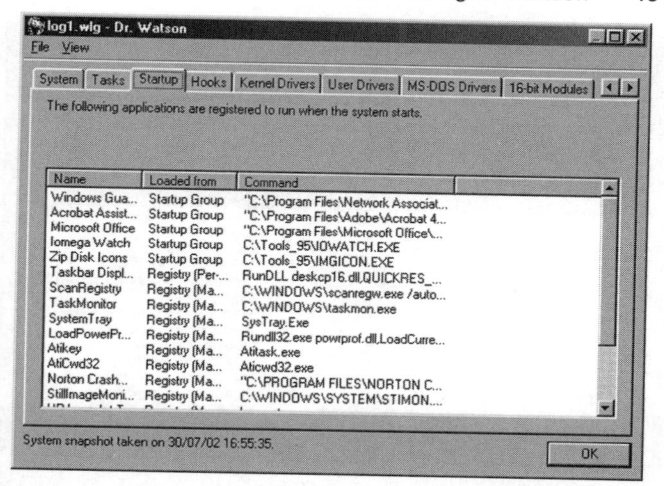

Figure 13.3 The Startup tab displays a list of the applications that are registered to run when the system starts. This information indicates whether the program is run from an entry in the Startup group or whether it is from the registry

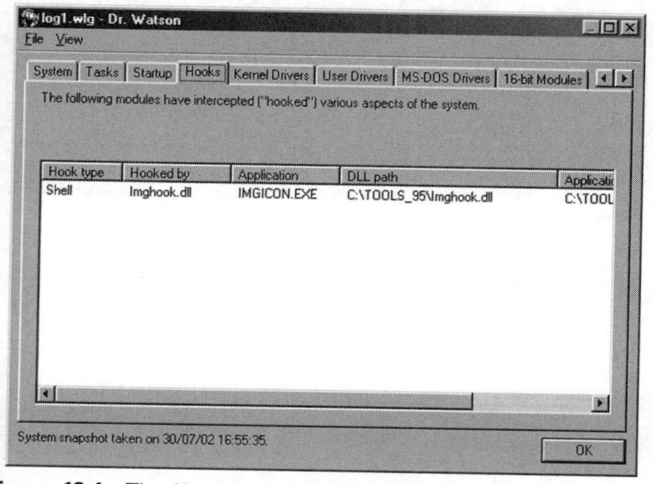

Figure 13.4 The Hooks tab provides information about modules that have intercepted (i.e. 'hooked') various aspects of the system. In this screen, Dr Watson is reporting a single hooked application, IMGICON.EXE

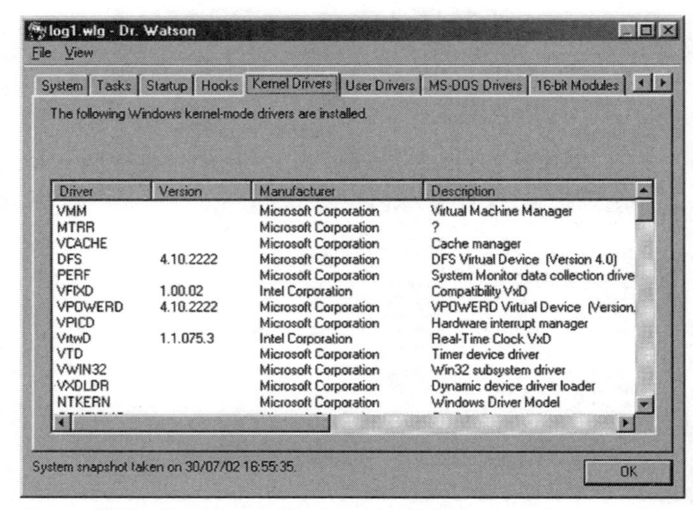

Figure 13.5 The Kernel Drivers tab displays a list of kernel mode drivers, including their manufacturer and version number

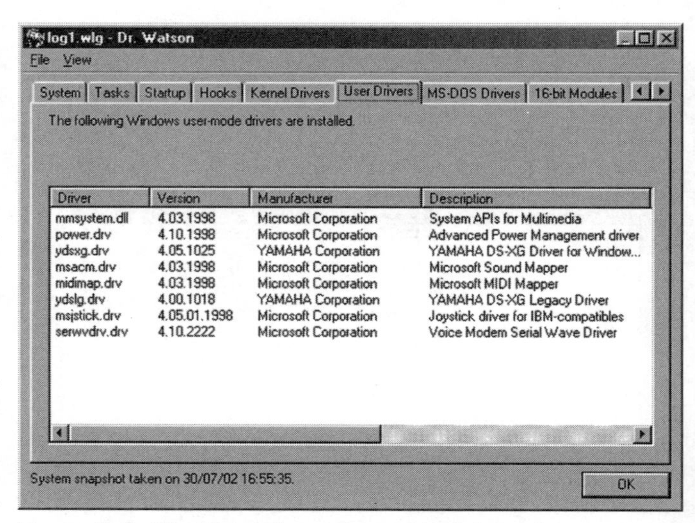

Figure 13.6 The User Drivers tab provides information on current user drivers. In this screen, Dr Watson is reporting on the various multimedia driver components. Once again, note the clarity and level of reporting provided by this excellent free tool

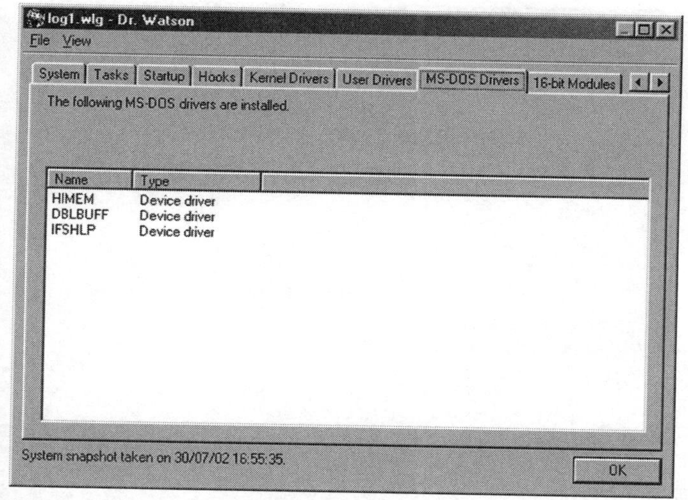

Figure 13.7 The MS-DOS Drivers tab reports on any MS-DOS drivers that happen to be present. These drivers are only used by DOS applications and not directly by Windows

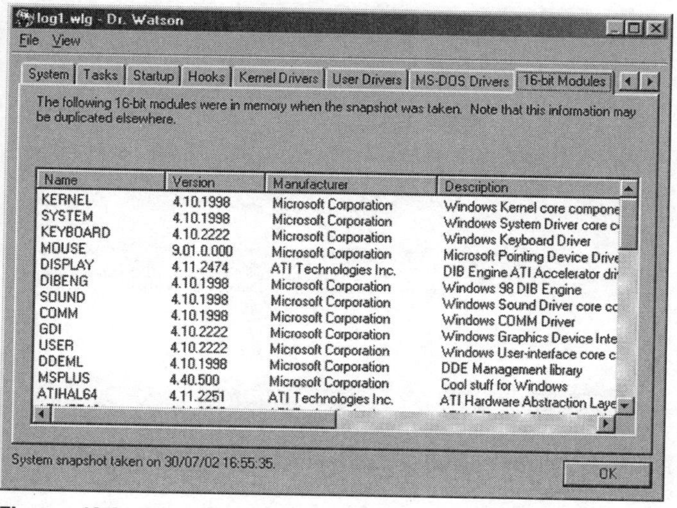

Figure 13.8 The 16-bit Modules tab provides information on Windows core components and modules such as the display driver

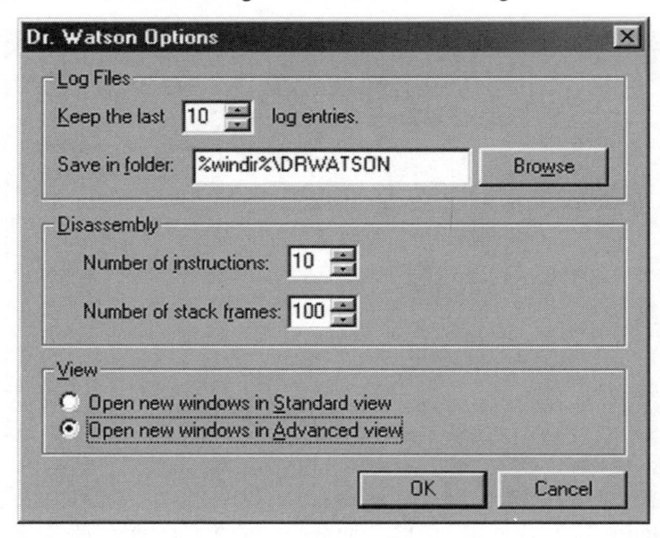

Figure 13.9 A limited range of configuration options is available within Dr Watson (see text)

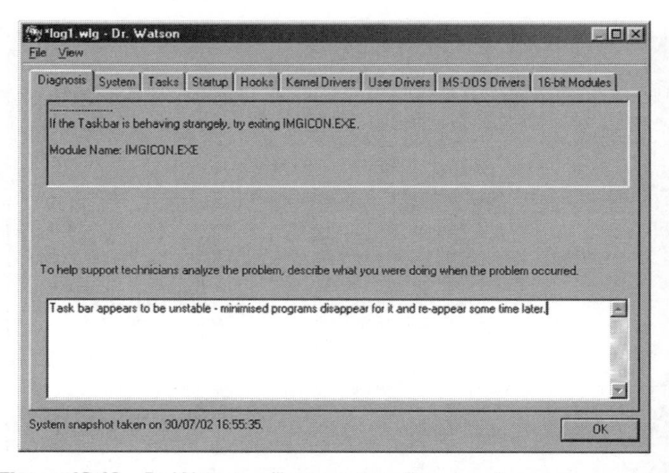

Figure 13.10 Dr Watson will attempt to diagnose the most recently occurring problem. Here, Dr Watson is suggesting that there might be an error related to the operation of the taskbar and has suggested that the executable file IMGICON.EXE might be at fault. The text field in the lower half of the tabbed window allows you to add comments that may prove to be helpful when any future faults arise

14 Troubleshooting Windows Registry

The current version of Windows on sale is XP but there are a significant number of machines being used with Windows 2000, 98 and ME. Unfortunately, these operating systems are so complex that several whole books would be required to describe how to troubleshoot them. Microsoft provide a large amount of troubleshooting information on their website at support.microsoft.com

This section of the book simply aims to provide you with some starting points for troubleshooting Windows.

All versions of Windows (except 3.1) use a large database of system information called the registry. This database is updated all the time, especially when software is installed. It stores settings, configurations and related data and is central to the way Windows works. If it becomes corrupted in any way, Windows will fail to load or run successfully. You should suspect registry trouble if Windows does not start after installing a new or updated piece of software.

The registry can be backed up and restored to overwrite a corrupted version, the method used to achieve this is different in the various versions of Windows.

14.1 Windows 95, 98 and ME

For detailed registry information, either use support.microsoft.com/default.aspx?scid=kb;en-us;322754 or search for 322754 in support.microsoft.com

To examine (and, if necessary, edit) the registry you can make use of the Windows *registry editor* (regedit.exe). This utility is in the \WINDOWS directory but you can run it by clicking start → run, enter the name regedit in the box and clicking OK. This should be approached with care as mistakes may require a complete reinstallation of Windows and all application software!

You can make a manual registry backup by following these steps:

Open a DOS window by following the sequence Start → Programs → MS-DOS prompt then type these commands followed by the *enter* key (also called *carriage return*)

```
cd \Windows
attrib -r -h -s system.dat
```

attrib -r -h -s user.dat
copy system.dat a:\system.dat
copy user.dat a:\user.dat

exit

The attrib commands change the *file attributes*, -r means 'not read only', -h means 'not hidden' and -s means 'not a system file'. These files are usually hidden, read-only files and Windows will reset the attributes on next bootup.

If you need to replace a corrupt registry from your backup, follow these commands:

Open a DOS window by following the sequence Start → Programs → MS-DOS prompt then type these commands followed by the *enter* key (also called *carriage return*)

From a previous copy on the system	From a copy on a floppy disk
cd \Windows	cd \Windows
attrib -r -h -s system.da0	attrib -r -h -s system.dat
attrib -r -h -s system.dat	copy a:\system.dat
ren system.dat system.old	attrib -r -h -s user.dat
ren system.da0 system.dat	copy a:\user.dat
attrib -r -h -s user.da0	exit
attrib -r -h -s user.dat	
ren user.dat user.old	
ren user.da0 user.dat	
exit	

Warning: If this fails, you will most likely have to re-install Windows and all your applications from scratch.

TIP: When running a DOS session in a window, you can toggle between the window and full-screen DOS by pressing <ALT> and <TAB> together.

TIP: Incorrect entries in the registry can cause unpredictable results (or Windows may simply refuse to run). You should therefore not make changes to the registry unless you are completely confident that you know what you are doing! Never make changes to the registry without first making a backup copy (see below).

TIP: The Windows 98 Registry Editor does not provide you with the registry data in the form of a single editable file. If you find this an inconvenient way of displaying/editing the registry files, you can use the Export Registry File option to create a text file containing the registry contents (the file created will have a .REG file extension). You can also import all (or part) of a modified registry data file. However, if you do this you must take great care to edit the file as an ASCII file and introduce no extra codes or unwanted formatting.

14.2 Windows 2000

To restore the registry in Windows 2000:

Click Start, click Shut Down, click Restart, and then click OK.
When the message appears 'Please select the operating system to start', press F8.
Then choose 'Last Known Good Configuration'.
Choose an operating system, and then press ENTER.

This method will allow recovery from the installation of a bad driver or incompatible piece of software.

For much more detailed information, either use support.microsoft.com/default.aspx?scid=kb;en-us;322755 or search for 322755 in support.microsoft.com

14.3 Windows XP

The registry in Windows XP is well protected so should not need to be manually edited; however, trouble can result from failed installations, bad or corrupted drivers, etc. Drivers for XP are 'digitally signed' to prove they have passed reliability tests set by Microsoft, so should not give trouble. The problem is that XP version drivers are not available for every device so attempts to use an older version may lead to trouble. The consequence of the XP version being unavailable results in useful hardware becoming redundant. For example, some versions of the Minolta Dimage Scan Dual film scanner cannot be used with XP as Minolta have not made the driver available, a device costing several hundred pounds must now be scrapped. Until this situation is resolved, it is best to use Windows 2000 where driver versions are far less of a problem.

For detailed registry information, either use support.microsoft.com/default.aspx?scid=kb;en-us;322756 or search for 322756 in support.microsoft.com

14.4 More links to registry related sites

14.4.1 Windows 95, 98 and ME

support.microsoft.com/default.aspx?scid=kb;en-us;151284

support.microsoft.com/default.aspx?scid=KB;en-us;q188867

(Windows 98 and ME) Description of Troubleshooting Settings for File System Properties
support.microsoft.com/default.aspx?scid=kb;en-us;247485

How to Perform Clean-Boot Troubleshooting for Windows 98
support.microsoft.com/default.aspx?scid=kb;en-us;192926

www.Windows-help.net/index.shtml

14.4.2 Windows 2000

www.labmice.net/troubleshooting/default.htm
www.labmice.net/troubleshooting/errorcodes.htm

14.4.3 Windows XP

www.labmice.net/WindowsXP/TroubleshootingXP/default.htm
www.annoyances.org/
from www.everythingcomputers.com/pc_startup_trouble_print.htm

14.4.4 General

www.Windowsreinstall.com/install/trouble/INDEX.HTM
www.Windowsgalore.com/

Appendix A Common file extensions

For a very much larger list, see

www.uni-koeln.de/themen/Graphik/ImageProcessing/fileext.html

Extension	Type of file
ASC	An ASCII text file
ASM	An assembly language source code file
BAK	A backup file (often created automatically by a text editor which renames the source file with this extension and the revised file assumes the original file specification)
BAS	A BASIC program source file
BAT	A batch file which contains a sequence of operating system commands
BIN	A binary file (comprising instructions and data in binary format)
BMP	A bit-mapped picture file
C	A source code file written in the C language
CLP	A Windows 'clipboard' file
COM	An executable program file in small memory format (i.e. confined to a single 64K byte memory segment)
CPI	A 'code page information' file
CRD	A Windows 'card index' file
DAT	A data file (usually presented in either binary or ASCII format)
DBG	A DEBUG text file
DOC	A document file (not necessarily presented in standard ASCII format)
EXE	An executable program file in large memory format (i.e. not confined to a 64K byte memory model)
GIF	A graphics image file
HEX	A file presented in hexadecimal (an intermediate format sometimes used for object code)
HTM	A file in HTML encoded format suitable for viewing in a web browser
INI	An initialization file which may contain a set of inference rules and/or environment variables
LIB	A library file (containing multiple object code files)

LST	A listing file (usually showing the assembly code corresponding to each source code instruction together with a complete list of symbols)
OBJ	An object code file
OLD	A backup file (replaced by a more recent version of the file)
PAS	A source code file written in Pascal
PCX	A picture file
PIF	A Windows 'program interchange file'
SCR	A DEBUG script file
SYS	A system file
TIF	A tagged image file
TMP	A temporary file
TXT	A text file (usually in ASCII format)
WRI	A document file produced by Windows 'Write'
$$$	A temporary file

Appendix B The Command Prompt and DOS

Note: Although DOS is little used in a world where Linux or Windows is common, this section has been left in the revised edition of the book because DOS can still provide some facilities that Windows cannot offer, for example writing a list of filenames into a file.

Some of the commands shown in this section will not work in a Windows DOS window or Command Prompt but have been left in for completeness.

The original PCs were supplied with DOS, the disk operating system. When Windows came out, DOS still formed the main core. This was true with all versions up to Windows 98. Later versions of Windows do not have DOS as the core but the command prompt still works in a very similar way. This chapter is designed to give you a good understanding of the resources provided by DOS.

B.1 I/O channels, DOS device names

In order to simplify the way in which DOS handles input and output, the system recognizes the names of its various I/O devices. This may at first appear to be unnecessarily cumbersome but it is instrumental in allowing DOS to redirect data. This feature can be extremely useful when, for example, output normally destined for the printer is to be redirected to an auxiliary serial port. See section B.3.3.

Table B.1 DOS devices

DOS device name		Usual name
CON:	Console	Keyboard
LPT1: LPT2: etc.	Line printer 1	Printer port
COM1: COM2: etc.	Communication port 1	Serial port
NUL:	Nothing!	
PRN:	Printer	Printer (serial or parallel)

B.2 DOS commands

DOS responds to command lines typed at the console and terminated with a <RETURN> or <ENTER> keystroke. A command line is thus composed of a command keyword, an optional command tail and <RETURN>. The command keyword identifies the command (or program) to be executed. The command tail can contain extra information relevant to the command, such as a filename or other parameters. Each command line must be terminated using <RETURN> or <ENTER> (not shown in the examples which follow).

As an example, the following command can be used to display a directory of all bit-mapped picture files (i.e. those with a BMP extension) within a directory named GALLERY in drive C:, indicating the size of each:

DIR C:\GALLERY*.BMP

Note that, in this example and the examples which follow, we have omitted the prompt generated by the system (indicating the current drive).

It should be noted that the command line can be entered in any combination of upper-case or lower-case characters. DOS converts all letters in the command line to upper-case before interpreting them. Furthermore, while a command line generally immediately follows the system prompt, DOS permits spaces between the prompt (e.g. C:\>) and the command word.

As characters are typed at the keyboard, the cursor moves to the right in order to indicate the position of the next character to be typed. Depending upon the keyboard used, a <BACKSPACE>, or <DELETE> key can be used to delete the last entered character and move the cursor backwards one character position. Alternatively, a combination of the CONTROL and H keys (i.e. <CTRL-H>) may be used instead.

TIP: <CTRL-ALT-DEL> can be used to perform a 'warm' system reset. This particular combination should only be used as a last resort as it will clear system memory. Any program or data present in RAM will be lost!

B.2.1 Repeating or editing DOS commands

If it is necessary to repeat or edit the previous command, the < F1 > (or right-arrow) key may be used to reproduce the command line, character by character, on the screen. The left-arrow key permits backwards movement through the command line for editing purposes. The < F3 > key simply repeats the last command in its entirety.

B.2.2 File specifications

Many of the DOS commands make explicit reference to files. A file is simply a collection of related information stored on a disk. Program files comprise a series of instructions to be executed by the processor whereas data files simply contain a collection of records. A complete file specification has four distinct parts; a drive and directory specifier (known as a 'pathname'), a filename and a filetype.

The drive specifier is a single letter followed by a colon (e.g. C:). This is then followed by the directory and sub-directory names (if applicable) and the filename and filetype. The filename comprises 1 to 8 characters while the filetype takes the form of a 1 to 3 character extension separated from the filename by means of a full stop ('.'). A complete file specification (or 'filespec') thus takes the form:

[pathname]:[filename].[filetype]

As an example, the following file specification refers to a file named MOUSE and having a COM filetype found in the root directory of the disk in drive A:

A\:MOUSE.COM

DOS allows files to be grouped together within directories and sub-directories. Directory and sub-directory names are separated by means of the backslash (\) character. Directories and sub-directories are organized in a hierarchical (tree) structure and thus complete file specifications must include directory information.

The 'root' or base directory (i.e. that which exists at the lowest level in the hierarchical structure) is accessed by default when we simply specify a drive name without further reference to a directory. Thus:

C:\MOUSE.COM

refers to a file in the root directory while:

C:\DOS\MOUSE.COM

refers to an identically named file resident in a sub-directory called 'DOS'.

Sub-directories can be extended to any practicable level. As an example:

C:\DOS\UTILS\MOUSE\MOUSE.COM

refers to a file named MOUSE.COM present in the MOUSE sub-directory which itself is contained within the UTILS sub-directory found within a directory named DOS.

When it is necessary to make explicit reference to the root directory, we can simply use a single backslash character as follows:

C:

B.2.3 File extensions

The filetype extension provides a convenient mechanism for distinguishing different types of file and DOS provides various methods for manipulating groups of files having the same filetype extension. We could, for example, delete all of the backup (BAK) present in the root directory of the hard disk (drive C:) using a single command of the form:

ERA C:*.BAK

Alternatively, we could copy all of the executable (EXE) files from the root directory of the disk in drive A: to the root directory on drive C: using the command:

COPY A:*.EXE C:

Commonly used filetype extensions are shown in Appendix A.

B.2.4 Wildcard characters

DOS allows the user to employ wildcard characters when specifying files. The characters '*' and '?' can be used to replace complete fields and individual characters respectively within a file specification. DOS will search then carry out the required operation on all files for which a match is obtained.

The following examples illustrate the use of wildcard characters:

A:*.COM

refers to all files having a COM extension present in the root directory of drive A:.

C:\TOOLS*.*

refers to all files (regardless of name or extension) present in the directory named TOOLS on drive C:.

B:\TURBO\PROG?.C

refers to all files having a C extension present in the TURBO directory on the disk in drive B which have PROG as their first three letters and any alphanumeric character in the fourth character place. A match will occur for each of the following files:

PROG1.C PROG2.C PROG3.C PROGA.C PROGB.C, etc.

B.3 Internal and external commands

It is worth making a distinction between DOS commands which form part of the resident portion of the operating system (internal commands) and those which involve other utility programs (external commands). Intrinsic commands are executed immediately whereas extrinsic commands require the loading of transient utility programs from disk and hence there is a short delay before the command is acted upon.

In the case of external commands, DOS checks only the command keyword. Any parameters which follow are passed to the utility program without checking.

At this point we should perhaps mention that DOS only recognizes command keywords which are correctly spelled! Even an obvious typing error will result in the non-acceptance of the command and the system will respond with an appropriate error message.

As an example, suppose you attempt to format a disk but type **FORMATT** instead of **FORMAT**. Your system will respond with this message:

Bad command or file name

indicating that the command is unknown and that no file of that name (with a COM, BAT, or EXE extension) is present in the current directory.

> **TIP:** To get on-line help from within MS-DOS 5.0 and DR-DOS 6.0 (and later operating systems) you can simply type the command name followed by **/?**. Hence **DIR/?** will bring you help before using the directory command. With MS-DOS 5.0 (and later) you can also type **HELP** followed by the command name (e.g. **HELP DIR**). In DR-DOS 6.0 you can type **DIR/H**.

B.3.1 Internal DOS commands

We shall now briefly examine the function of each of the most commonly used internal DOS commands. Examples have been included wherever they can help to clarify the action of a particular command. The examples relate to the most commonly used versions of MS-DOS, PC-DOS, and DR-DOS.

Command	Function
BREAK	The **BREAK** command disables the means by which it is possible to abort a running program. This facility is provided by means of the <CTRL-C> or <CTRL-BREAK> key combinations and it normally only occurs when output is being directed to the screen or the printer. **BREAK** accepts two parameters, **ON** and **OFF**. Examples: **BREAK ON** enables full <CTRL-C> or <CTRL-BREAK> key checking (it is important to note that this will normally produce a dramatic reduction in the speed of execution of a program). **BREAK OFF** restores normal <CTRL-C> or <CTRL-BREAK> operation (i.e. the default condition).

> **TIP: BREAK ON** will often result in a significant reduction in the speed of execution of a program. You should only use this command when strictly necessary!

Command	Function
CD	See **CHDIR**.
CHDIR	The **CHDIR** command allows users to display or change the current directory. **CHDIR** may be abbreviated to **CD**. Examples: **CHDIR A:** displays the current directory path for the disk in drive A:. **CHDIR C:\APPS** changes the directory path to APPS on drive C:.

CD D:\DEV\PROCESS

changes the directory path to the sub-directory PROCESS within the directory named DEV on drive D:.

CD

changes the directory path to the root directory of the *current* drive.

CD..

changes the directory path one level back towards the root directory of the *current* drive.

CLS

CLS clears the screen and restores the cursor position to the top left-hand corner of the screen.

COPY

The **COPY** command can be used to transfer a file from one disk to another using the same or a different filename. The **COPY** command is effective when the user has only a single drive. The **COPY** command must be followed by one or two file specifications. When only a single file specification is given, the command makes a single drive copy of a file. The copied file takes the same filename as the original and the user is prompted to insert the source and destination disks at the appropriate point. Where both source and destination file specifications are included, the file is copied to the specified drive and the copy takes the specified name. Where only a destination drive is specified (i.e. the destination filename is omitted) the **COPY** command copies the file to the specified drive without altering the filename. **COPY** may be used with the ***** and **?** wildcard characters in order to copy all files for which a match is found.
Examples:

COPY A\:ED.COM B:

copies the file ED.COM present in the root directory of the disk in drive A: to the disk present in drive B:. The copy will be given the name ED.COM.

TIP: On a single drive system the only available floppy drive can be used as both the source and destination when the **COPY** command is used. The single physical drive will operate as both drive A: and drive B: and you will be prompted to insert the source and destination disks when required.

TIP: COPY is unable to make copies of files located within sub-directories. If you need this facility use **XCOPY** with the **/s** switch.

DATE

The **DATE** command allows the date to be set or displayed.
Examples:

DATE

displays the date on the screen and also prompts the user to make any desired changes. The user may press <RETURN> to leave the settings unchanged.

DATE 12-27-99

sets the date to 27 December 1999.

DEL

See **ERASE**.

DIR

The **DIR** command displays the names of files present within a directory. Variations of the command allow the user to specify the drive to be searched and the types of files to be displayed. Further options govern the format of the directory display.
Examples:

DIR

displays all files in the current default directory.

A:\ DIR

changes the default drive to A: (root directory) and then displays the contents of the root directory of the disk in drive A:.

DIR *.BAS

displays all files with a BAS extension present in the current default directory drive.

DIR C:\DEV.*

displays all files named DEV (regardless of their type or extension) present in the root directory of drive C: (the hard disk).

DIR C:\MC*.BIN

displays all files having a BIN extension present in the sub-directory named MC on drive C: (the hard disk).

DIR/W

displays a directory listing in 'wide' format (excluding size and creation date/time information) of the current default directory.

TIP: To prevent directory listings scrolling off the screen use **DIR /P** or **DIR | MORE**. These commands will pause the listing at the end of each screen and wait for you to press a key before continuing.

> **TIP:** MS-DOS 5.0 (and later) includes many options for use with the **DIR** command including sorting the directory listing and displaying hidden system files.

ERASE

The **ERASE** command is used to erase a filename from the directory and release the storage space occupied by a file. The **ERASE** command is identical to the **DEL** command and the two may be used interchangeably. **ERASE** may be used with the * and ? wildcard characters in order to erase all files for which a match occurs.
Examples:

ERASE PROG1.ASM

erases the file named PROG1.ASM from the disk placed in the current (default) directory.

ERASE B:\TEMP.DAT

erases the file named TEMP.DAT from the root directory of the disk in drive B:.

ERASE C:*.COM

erases all files having a COM extension present in the root directory of the hard disk (drive C:).

ERASE A:\PROG1.*

erases all files named PROG1 (regardless of their type extension) present in the root directory of the disk currently in drive A:.

MD

See **MKDIR**.

MKDIR

The **MKDIR** command is used to make a new directory or sub-directory. The command may be abbreviated to **MD**.
Examples:

MKDIR APPS

creates a sub-directory named APPS within the *current* directory (note that the **CHDIR** command is often used after **MKDIR** – having created a new directory you will probably want to make it the current directory before doing something with it!).

MD C:\DOS\BACKUP

creates a sub-directory named BACKUP within the DOS directory of drive C:.

PATH

The **PATH** command may be used to display the current directory path. Alternatively, a new directory path may be established using the **SET PATH** command.
Examples:

PATH

displays the current directory path (a typical response would be **PATH=C:\WINDOWS**).

SET PATH=C:\DOS

makes the directory path C:\DOS.

PROMPT

The **PROMPT** command allows the user to change the system prompt. The **PROMPT** command is followed by a text string which replaces the system prompt. Special characters may be inserted within the string, as follows:

$d current date
$e escape character
$g >
$h backspace and erase
$l <
$n current drive
$p current directory path
$q =
$t current time
$v DOS version number
$$ $
$ newline

Examples:

PROMPT tg
changes the prompt to the current time followed by a >.

PROMPT Howard Associates PLC $?

changes the prompt to Howard Associates PLC followed by a carriage return and newline on which a ? is displayed.

PROMPT

restores the default system prompt (e.g. C:\>).

> **TIP:** The most usual version of the **PROMPT** command is **PROMPT pg** which displays the current directory/sub-directory and helps to avoid confusion when navigating within DOS directories.

RD

See **RMDIR**.

RENAME

The **RENAME** command allows the user to rename a disk file.
RENAME may be used with the * and **?** wildcard characters in order to rename all files for which a match occurs. **RENAME** may be abbreviated to **REN**.
Examples:

RENAME PROG2.ASM PROG1.ASM

renames PROG1.ASM to PROG2.ASM on the disk placed in the current (default) directory.

REN A:\HELP.DOC HELP.TXT

renames the file HELP.DOC to HELP.TXT in the root directory of the disk in drive A:.

REN B:\CONTROL.* PROG1.*

renames all files with the name PROG1 (regardless of type extension) to CONTROL (with identical extensions) found in the root directory of the disk in drive B:.

RMDIR

The **RMDIR** command is used to remove a directory. **RMDIR** may be abbreviated to **RD**. The command cannot be used to remove the current directory and any directory to be removed must be empty and must not contain further sub-directories.
Example:

RMDIR ASSEM

removes the directory ASSEM from the current directory (note that DOS will warn you if the named directory is *not* empty!).

RD C:\DOS\BACKUP

removes the directory ASSEM from the current directory (once again, DOS will warn you if the named directory is *not* empty!).

SET

The **SET** command is used to set the environment variables (see **PATH**).

TIME

The **TIME** command allows the time to be set or displayed.
Examples:

TIME

displays the time on the screen and also prompts the user to make any desired changes. The user may press <RETURN> to leave the settings unchanged.

TIME 14:30

sets the time to 2.30 p.m.

TYPE

This useful command allows you to display the contents of an ASCII (text) file on the console screen. The **TYPE** command can be used with options which enable or disable paged mode displays. The <PAUSE> key or <CTRL-S> combination may be used to halt the display. You can press any key or use the <CTRL-Q> combination respectively to restart. <CTRL-C> may be used to abort the execution of the **TYPE** command and exit to the system.
Examples:

TYPE C\:AUTOEXEC.BAT

will display the contents of the AUOTEXEC.BAT file stored in the root directory of drive C:. The file will be sent to the screen.

TYPE B\:PROG1.ASM

will display the contents of a file called PROG1.ASM stored in the root directory of the disk in drive B. The file will be sent to the screen.

TYPE C:\WORK*.DOC

will display the contents of *all* the files with a DOC extension present in the WORK directory of the hard disk (drive C:).

TIP: You can use the **TYPE** command to send the contents of a file to the printer at the same time as viewing it on the screen. If you need to do this, press <CTRL-P> before you issue the **TYPE** command (but do make sure that the printer is 'on-line' and ready to go!). To disable the printer output you can use the <CTRL-P> combination a second time.

TIP: The ability to redirect data is an extremely useful facility. DOS uses the < and > characters in conjunction with certain commands to redirect files. As an example:

TYPE A:\README.DOC >PRN

will redirect normal screen output produced by the TYPE command to the printer. This is usually more satisfactory than using the <PRT.SCREEN> key.

VER	The **VER** command displays the current DOS version.
VERIFY	The **VERIFY** command can be used to enable or disable disk file verification. **VERIFY ON** enables verification while **VERIFY OFF** disables verification. If **VERIFY** is used without **ON** or **OFF**, the system will display the state of verification (either 'on' or 'off').
VOL	The **VOL** command may be used to display the volume label of a disk.

B.3.2 External DOS commands

Unlike internal commands, these commands will not function unless the appropriate DOS utility program is resident in the current (default) directory. External commands are simply the names of utility programs (normally resident in the DOS sub-directory). If you need to gain access to these utilities from any directory or sub-directory, then the following lines should be included in your AUTOEXEC.BAT file:

SET PATH=C:\DOS

The foregoing assumes that you have created a sub-directory called DOS on the hard disk and that this sub-directory contains the DOS utility programs. As with the internal DOS commands, the examples given apply to the majority of DOS versions.

Command	Function
APPEND	The **APPEND** command allows the user to specify drives, directories and sub-directories which will be searched through when a reference is made to a particular data file. The **APPEND** command follows the same syntax as the **PATH** command.
ASSIGN	The **ASSIGN** command allows users to redirect files between drives. **ASSIGN** is particularly useful when a RAM disk is used to replace a conventional disk drive. Examples: **ASSIGN A=D** results in drive D: being searched for a file whenever a reference is made to drive A:. The command may be countermanded by issuing a command of the form: **ASSIGN A=A**

Alternatively, all current drive assignments may be overridden by simply using:

ASSIGN

TIP: ASSIGN A=B followed by **ASSIGN B=A** can be used to swap the drives over in a system that has two floppy drives. The original drive assignment can be restored using **ASSIGN**.

ATTRIB

The **ATTRIB** command allows the user to examine and/or set the attributes of a single file or a group of files. The **ATTRIB** command alters the file attribute byte (which appears within a disk directory) and which determines the status of the file (e.g. read-only).
Examples:

ATTRIB A:\PROCESS.DOC

displays the attribute status of copies the file PROCESS.DOC contained in the root directory of the disk in drive A:.

ATTRIB +R A:\PROCESS.DOC

changes the status of the file PROCESS.DOC contained in the root directory of the disk in drive A: so that is a read-only file. This command may be countermanded by issuing a command of the form:

ATTRIB -R A:\PROCESS.DOC

TIP: A crude but effective alternative to password protection is that of using **ATTRIB** to make all the files within a sub-directory hidden. As an example, **ATTRIB +H C:\PERSONAL** will hide all of the files in the PERSONAL sub-directory. **ATTRIB -H C:\PERSONAL** will make them visible once again.

BACKUP

The **BACKUP** command may be used to copy one or more files present on a hard disk to a number of floppy disks for security purposes. It is important to note that the **BACKUP** command stores files in a compressed format (i.e. not in the same format as that used by the **COPY** command). The **BACKUP** command may be used selectively with various options including those which allow files to be archived by date. The **BACKUP** command usually requires that the target disks have been previously formatted; however, from MS-DOS 3.3 onwards, an option to format disks has been included.

Examples:

BACKUP C:*.* A:

backs up all of the files present on the hard disk. This command usually requires that a large number of (formatted) disks are available for use in drive A:. Disks should be numbered so that the data can later be restored in the correct sequence.

BACKUP C:\DEV*.C A:

backs up all of the files with a C: extension present within the DEV sub-directory on drive C:.

BACKUP C:\PROCESS*.BAS A:/D:01-01-99

backs up all of the files with a BAS extension present within the PROCESS sub-directory of drive C: that were created or altered on or after 1 January 1999.

BACKUP C:\COMMS*.* A:/F

backs up all of the files present in the COMMS sub-directory of drive C: and formats each disk as it is used.

CHKDSK

The **CHKDSK** command reports on disk utilization and provides information on total disk space, hidden files, directories and user files. **CHKDSK** also gives the total memory and free memory available. **CHKDSK** incorporates options which can be used to enable reporting and to repair damaged files. **CHKDSK** provides two useful switches; **/F** fixes errors on the disk and **/V** displays the name of each file in every directory as the disk is checked. Note that if you use the **/F** switch, **CHKDSK** will ask you to confirm that you actually wish to make changes to the disk's file allocation table (FAT).
Examples:

CHKDSK A:

checks the disk placed in the A: drive and displays a status report on the screen.

CHKDSK C:\DEV*.ASM/F/V

checks the specified disk and directory, examining all files with an ASM extension, reporting errors and attempting to correct them.

TIP: If you make use of the **/F** switch, **CHKDSK** will ask you to confirm that you actually wish to correct the errors. If you do go ahead **CHKDSK** will usually change the disk's file allocation table (FAT). In some cases this may result in loss of data!

TIP: The CHKDSK command has a nasty bug in certain versions of MS-DOS and PC-DOS. The affected versions are:

DOS version	File name	File size	Data
PC-DOS 4.01	CHKDSK.COM	17771 bytes	17 Jun 88
MS-DOS 4.01	CHKDSK.COM	17787 bytes	30 Nov 88
PC-DOS 5.0	CHKDSK.EXE	16200 bytes	09 Apr 91
MS-DOS 5.0	CHKDSK.EXE	16184 bytes	09 May 91

The bug destroys the directory structure when CHKDSK is used with the /F switch and the total allocation units on disk is greater than 65 278. The bug was corrected in maintenance release 5.0A dated 11 November 91; however, the problem does not arise if the hard disk partition is less than 128 Mbytes.

If you have an affected DOS version it is well worth upgrading to avoid the disastrous consequences of this bug!

COMP

The **COMP** command may be used to compare two files on a line-by-line or character-by-character basis. The following options are available:

/A use . . . to indicate differences
/B perform comparison on a character basis
/C do not report character differences
/L perform line comparison for program files
/N add line numbers
/T leave tab characters
/W ignore white space at beginning and end of lines

Example:

COMP /B PROC1.ASM PROC2.ASM

carries out a comparison of the files PROC1.ASM and PROC2.ASM on a character-by-character basis.

DISKCOMP

The **DISKCOMP** command provides a means of comparing two (floppy) disks. **DISKCOMP** accepts drive names as parameters and the necessary prompts are generated when a single-drive disk comparison is made.
Example:

DISKCOMP A: B:

compares the disk in drive A: with that placed in drive B:.

EXE2BIN

The **EXE2BIN** utility converts, where possible, an EXE program file to a COM program file (which loads faster and makes less demands on memory space).
Example:

EXE2BIN PROCESS

will search for the program PROCESS.EXE and generate a program PROCESS.COM.

> **TIP: EXE2BIN** will not operate on EXE files that require more than 64K bytes of memory (including space for the stack and data storage) and/or those that make reference to other memory segments (CS, DS, ES and SS *must* all remain the same during program execution).

FASTOPEN

The **FASTOPEN** command provides a means of rapidly accessing files. The command is only effective when a hard disk is fitted and should ideally be used when the system is initialized (e.g. from within the AUTOEXEC.BAT file).
Example:

FASTOPEN C:32

enables fast opening of files and provides for the details of up to 32 files to be retained in RAM.

> **TIP: FASTOPEN** retains details of files within RAM and must not be used concurrently with **ASSIGN**, **JOIN** and **SUBST**.

FDISK

The **FDISK** utility allows users to format a hard (fixed) disk. Since the command will render any existing data stored on the disk inaccessible, **FDISK** should be used with extreme caution. Furthermore, improved hard disk partitioning and formatting utilities are normally supplied when a hard disk is purchased. These should be used in preference to **FDISK** whenever possible.

> **TIP:** To ensure that FDISK is not used in error, copy FDISK to a sub-directory that is not included in the PATH statement then erase the original version using the following commands:

```
CD\
MD XDOS
COPY C:\DOS\FDISK.COM C:\XDOS
ERASE C:\DOS\FDISK.COM
```

> Finally, create a batch file, FDISK.BAT, along the following lines and place it in the DOS directory:

```
ECHO OFF
CLS
ECHO ***** You are about to format the hard disk! *****
ECHO All data will be lost – if you do wish to continue
ECHO change to the XDOS directory and type FDISK again.
```

FIND

The **FIND** command can be used to search for a character string within a file. Options include:

/C display the line number(s) where the search string has been located
/N number the lines to show the position within the file
/V display all lines which do not contain the search string

Example:

FIND/C "output" C:/DEV/PROCESS.C

searches the file PROCESS.C present in the DEV sub-directory for occurrences of output. When the search string is located, the command displays the appropriate line number.

FORMAT

The **FORMAT** command is used to initialize a floppy or hard disk. The command should be used with caution since it will generally not be possible to recover any data which was previously present. Various options are available including:

/1 single-sided format
/8 format with 8 sectors per track
/B leave space for system tracks to be added (using the **SYS** command)
/N:8 format with 8 sectors per track
/S write system tracks during formatting (note that this must be the last option specified when more than one option is required)
/T:80 format with 80 tracks
/V format and then prompt for a volume label

Examples:

FORMAT A:

formats the disk placed in drive A:.

FORMAT B:/S

formats the disk placed in drive B: as a system disk

TIP: When you format a disk using the **/S** option there will be less space on the disk for user programs and data. As an example, the system files for DR-DOS 6.0 consume over 100 Kbytes of disk space!

JOIN

The **JOIN** command provides a means of associating a drive with a particular directory path. The command must be used with care and must not be used with **ASSIGN**, **BACKUP**, **DISKCOPY**, **FORMAT**, etc.

KEYB

The **KEYB** command invokes the DOS keyboard driver. **KEYB** replaces earlier utilities (such as **KEYBUK**) which were provided with DOS versions prior to MS-DOS 3.3. The command is usually incorporated in an AUTOEXEC.BAT file and must specify the country letters required.
Example:

KEYB UK

selects the UK keyboard layout.

LABEL

The **LABEL** command allows a volume label (maximum 11 characters) to be placed in the disk directory.
Example:

LABEL A: TOOLS

will label the disk present in drive A: as TOOLS. This label will subsequently appear when the directory is displayed.

MODE

The **MODE** command can be used to select a range of screen and printer options. **MODE** is an extremely versatile command and offers a wide variety of options.
Examples:

MODE LPT1: 120,6

initializes the parallel printer LPT1 for printing 120 columns at 6 lines per inch.

MODE LPT2: 60,8

initializes the parallel printer LPT2 for printing 60 columns at 8 lines per inch.

MODE COM1: 1200,N,8,1

initializes the COM1 serial port for 1200 baud operation with no parity, 8 data bits and 1 stop bit.

MODE COM2: 9600,N,7,2

initializes the COM2 serial port for 9600 baud operation with no parity, 7 data bits and 2 stop bits.

MODE 40

sets the screen to 40 column text mode.

MODE 80

sets the screen to 80 column mode.

MODE BW80

sets the screen to monochrome 40 column text mode.

MODE CO80

sets the screen to colour 80 column mode.

MODE CON CODEPAGE PREPARE=((850)C:\DOS\EGA.CPI)

loads codepage 850 into memory from the file EGA.CPI located within the DOS directory.

> **TIP:** The **MODE** command can be used to redirect printer output from the parallel port to the serial port using **MODE LPT1:=COM1:**. Normal operation can be restored using **MODE LPT1:**.

PRINT

The **PRINT** command sends the contents of an ASCII text file to the printer. Printing is carried out as a background operation and data is buffered in memory. The default buffer size is 512 bytes; however, the size of the buffer can be specified using **/B:** (followed by required buffer size in bytes). When the utility is first entered, the user is presented with the opportunity to redirect printing to the serial port (COM1:). A list of files (held in a queue) can also be specified.
Examples:

PRINT README.DOC

prints the file README.DOC from the current directory.

PRINT /B:4096 HELP1.TXT HELP2.TXT HELP3.TXT

establishes a print queue with the files HELP1.TXT HELP2.TXT, and HELP3.TXT and also sets the print buffer to 4 Kbytes. The files are sent to the printer in the specified sequence.

RESTORE

The **RESTORE** command is used to replace files on the hard disk which were previously saved on floppy disk(s) using the **BACKUP** command. Various options are provided (including restoration of files created before or after a specified date.
Examples:

RESTORE C:\DEV\PROCESS.COM

restores the files PROCESS.COM in the sub-directory named DEV on the hard disk partition, C:. The user is prompted to insert the appropriate floppy disk (in drive A:).

RESTORE C:\BASIC /M

restores all modified (altered or deleted) files present in the sub-directory named BASIC on the hard disk partition, C:.

SYS

The **SYS** command creates a new boot disk by copying the hidden DOS system files. **SYS** is normally used to transfer system files to a disk which has been formatted with the **/S** or **/B** option. **SYS** cannot be used on a disk which has had data written to it after initial formatting.

TREE

The **TREE** command may be used to display a complete directory listing for a given drive. The listing starts with the root directory.

XCOPY

The **XCOPY** utility provides a means of selectively copying files. The utility creates a copy which has the same directory structure as the original. Various options are provided:

- **/A** only copy files which have their archive bit set (but do not reset the archive bits)
- **/D** only copy files which have been created (or that have been changed) after the specified date
- **/M** copy files which have their archive bit set but reset the archive bits (to avoid copying files unnecessarily at a later date)
- **/P** prompt for confirmation of each copy
- **/S** copy files from sub-directories
- **/V** verify each copy
- **/W** prompt for disk swaps when using a single drive machine

Example:

XCOPY C:\DOCS*.* A:/M

copies all files present in the DOCS sub-directory of drive C:. Files will be copied to the disk in drive A:. Only those files which have been modified (i.e. had their archive bits set) will be copied.

TIP: Always use XCOPY in preference to COPY when sub-directories exist. As an example, XCOPY C:\DOS*.* A:\ /S will copy all files present in the DOS directory on drive C: together with all files present in any sub-directories, to the root directory of the disk in A:.

B.3.3 Pipes, filters and IO redirection

B.3.3.1 I/O redirection

I/O means input output. It may come as a surprise but DOS does not show the result of commands on the screen nor does it read from the keyboard! You might argue that you can see the output but in fact DOS writes to something called STDOUT (Standard Output) which is the screen *unless you tell it otherwise*. It reads from STDIN (Standard Input) which happens to be the keyboard *unless you tell it otherwise*. All errors are sent to STDERR which is usually the screen.

The dir command is *redirected* into a file with the command

dir > dirlist.txt

The > character *redirects* the output from STDOUT to the file called dirlist.txt. The result is that you will not see the dir listing on the screen as usual, it will be in a text file that was created by the command. If the file had been there before, the old contents will now be lost and overwritten by the new dir listing details. If you did not want this to happen, you can add the results to the end of the file with the command

dir > > dirlist.txt

where the > > characters mean 'append to' following the redirection.

Question

How would you make a file called football.txt that contains

- A list of all files that start with the word 'foot'
- Files to be in date order
- Files to be from all subdirectories on the disk.

Answer

Use the command

dir \foot*.* /s > football.txt

The part that says dir \foot*.* will search from \, the root directory and look for all files that start with foot but end with anything and have any extension. The /s is called switch s and makes the dir command look in all sub-directories. Finally the > redirects the output to the file called football.txt.

Question

How would you make a file called mywork.txt containing all the names of the files on a floppy disk in size order.

Answer

Use the command

DIR A:\ /S > MYFILE.TXT

It is interesting to note that Windows will not do this simple service but in old fashioned DOS it is very simple. Once the file is created you can load it into any editor you wish, even Word.

B.3.3.2 Pipes

A pipe in the language of operating systems is a means of connecting the output of one program to another. DOS contains a command called *sort* which, as the name suggests, will sort a list of words. You can use the sort command from the DOS prompt but it is not very useful. Try this

SORT (and press ENTER)
(type a few lines of words such as:)
burt
anne
joe
jim
fred
(then type CTRL Z, the ASCII End of File character)

The result should be

anne
burt
fred
jim
joe

i.e. all the names in order.

The real use of this program is as a 'filter' when a command is 'piped' to the sort program.

Try this command:

DIR /A-D | SORT

the dir /a-d command will work as above but the result will be *piped* into the sort command with the | character. This is called the pipe character

and is usually above the \ character to the left of the z key. NB. *In a DOS Window, some keyboard maps do not yield the same characters as are marked on the keys. The pipe character is ASCII 124 and can be obtained by holding down the ALT key, keying 124 on the* numeric keypad *then releasing the ALT key. If you try the command below and the system responds with the message 'Too many parameters Find' it is because you have the wrong character used for the | symbol.*
The command

DIR /A-D | FIND "FOOT"

will send the output of the dir command into the *find* program (a filter program) and look for all occurrences of the string 'foot' in the listing. Try the command for strings that you know are present.

The DOS wildcard does not work correctly. If you want all files that contain the string 'fred', the command dir *fred.* should work but does not, it gives all the files on the machine. You can get the result you want by using

DIR | FIND "FRED"

because the output of the command is piped to the find program which looks for strings in lines of text.

This method can be very useful if you need to find a file made on a given date but you cannot remember the name. The command

DIR /S | FIND "14/09/98"

will look in all subdirectories and then search in each line for the string '14/9/98'. You need care here because DOS is capable of displaying the date in different formats. You may find the date 14 September 1998 looks like 14/9/98 or 9/14/98 in US format or 14-09-98, etc.

Of course if you use the command

DIR /S | FIND "14/09/98" > MYLIST.TXT

the output will be sent to a file called mylist.txt.

If you are feeling adventurous, you can combine all these things. A command such as:

DIR C:*SYS*.* /S | FIND "98" | SORT > C:\THIS YEAR.TXT

will make a file, in alphabetical order, of all files that contain the string 'sys' in the name and the string '98' somewhere in the dir line.

Appendix C Using batch files

Batch files provide a means of avoiding the tedium of repeating a sequence of operating system commands many times over. Batch files are nothing more than straightforward ASCII text files which contain the commands which are to be executed when the name of the batch is entered. Execution of a batch file is automatic; the commands are executed just as if they had been typed in at the keyboard. Batch files may also contain the names of executable program files (i.e. those with a COM or EXE extension), in which case the specified program is executed and, provided the program makes a conventional exit to DOS upon termination, execution of the batch file will resume upon termination.

C.1 Batch file commands

DOS provides a number of commands which are specifically intended for inclusion within batch files.

Command	Function
ECHO	The **ECHO** command may be used to control screen output during execution of a batch file. **ECHO** may be followed by **ON** or **OFF** or by a text string which will be displayed when the command line is executed. Examples: **ECHO OFF** disables the echoing (to the screen) of commands contained within the batch file. **ECHO ON** re-enables the echoing (to the screen) of commands contained within the batch file. (Note that there is no need to use this command at the end of a batch file as the reinstatement of screen echo of keyboard generated commands is automatic). **ECHO Sorting data – please wait!** displays the message: **Sorting data – please wait!** on the screen.

TIP: You can use **@ECHO OFF** to disable printing of the ECHO command itself. You will normally want to use this command instead of **ECHO OFF**.

FOR

FOR is used with **IN** and **DO** to implement a series of repeated commands.
Examples:

FOR %A IN (IN.DOC OUT.DOC MAIN.DOC) DO COPY %A LPT1:

copies the files IN.DOC, OUT.DOC and MAIN.DOC in the current directory to the printer.

FOR %A IN (*.DOC) DO COPY %A LPT1:

copies all the files having a DOC extension in the current directory to the printer. The command has the same effect as **COPY *.DOC LPT1:**.

IF

IF is used with **GOTO** to provide a means of branching within a batch file. **GOTO** must be followed by a label (which must begin with :).
Example:

IF NOT EXIST SYSTEM.INI GOTO :EXIT

transfers control to the label **:EXIT** if the file SYSTEM.INI cannot be found in the current directory.

PAUSE

the pause command suspends execution of a batch file until the user presses any key. The message:

Press any key when ready...

is displayed on the screen.

REM

The **REM** command is used to precede lines of text which will constitute remarks.
Example:

REM Check that the file exists before copying

C.2 Creating batch files

Batch files may be created using an ASCII text editor or a word processor (operating in ASCII mode). Alternatively, if the batch file comprises only a few lines, the file may be created using the DOS COPY command. As an example, let us suppose that we wish to create a batch file which will:

1. Erase all of the files present on the disk placed in drive B:.
2. Copy all of the files in drive A having a TXT extension to produce an identically named set of files on the disk placed in drive B:.
3. Rename all of the files having a TXT extension in drive A: so that they have a BAK extension.

The required operating system commands are thus:

ERASE B:*.*
COPY A:*.TXT B:
RENAME A:*.TXT A:*.BAK

The following keystrokes may be used to create a batch file named ARCHIVE.BAT containing the above commands (note that <ENTER> is used to terminate each line of input):

COPY CON: ARCHIVE.BAT
ERASE B:*.*
COPY A:*.TXT B:
RENAME A:*.TXT A:*.BAK
<CTRL-Z>

If you wish to view the batch file which you have just created simply enter the command:

TYPE ARCHIVE.BAT

Whenever you wish to execute the batch file simply type:

ARCHIVE

Note that, if necessary, the sequence of commands contained within a batch file may be interrupted by typing:

<CTRL-C>

(i.e. press and hold down the CTRL key and then press the C key).

The system will respond by asking you to confirm that you wish to terminate the batch job. Respond with **Y** to terminate the batch process or **N** if you wish to continue with it.

Additional commands can be easily appended to an existing batch file. Assume that we wish to view the directory of the disk in drive A after running the archive batch file. We can simply append the extra commands to the batch files by entering:

COPY ARCHIVE.BAT + CON:

The system displays the filename followed by the CON prompt. The extra line of text can now be entered using the following keystrokes (again with each line terminated by <ENTER>):

DIR A:
<CTRL-Z>

> **TIP:** Although you can use the COPY CON technique to create batch files, it is easier to use a text editor. If you must, you can use Windows Notepad but the best editor by far is called Ultraedit. See www.ultraedit.com

C.3 Passing parameters

Parameters may be passed to batch files by including the % character to act as a place holder for each parameter passed. The parameters are numbered strictly in the sequence in which they appear after the name of the batch file. As an example, suppose that we have created a batch file called REBUILD, and this file requires two file specifications to be passed as parameters. Within the text of the batch file, these parameters will be represented by %1 and %2. The first file specification following the name of the batch file will be %1 and the second will be %2. Hence, if we enter the command:

REBUILD PROC1.DAT PROC2.DAT

During execution of the batch file, %1 will be replaced by PROC1.DAT whilst %2 will be replaced by PROC2.DAT.

It is also possible to implement simple testing and branching within a batch file. Labels used for branching should preferably be stated in lower case (to avoid confusion with operating systems commands) and should be preceded by a colon when they are the first (or only) statement in a line. The following example which produces a sorted list of directories illustrates these points:

```
@ECHO OFF
IF EXIST %1 GOTO valid
ECHO Missing or invalid parameter
GOTO end
:valid
ECHO Index of Directories in %1
DIR %1 | FIND "<DIR>" | SORT
:end
```

The first line disables the echoing of subsequent commands contained within the batch file. The second line determines whether, or not, a valid parameter has been entered. If the parameter is invalid (or missing) the **ECHO** command is used to print an error message on the screen.

Appendix D Using DEBUG

Debug is a very odd program! It can be used to look into RAM directly or into any file, it can assemble or unassemble files, it can run programs and be used to change as little as 1 bit in a file. **Use it with great care!**

It is started at the DOS prompt by

C:\ > DEBUG yourfile.txt

and all you get is a '-' character!

This '-' character is the input prompt. Possible commands are shown below:

```
assemble          A [address]
compare           C range address
dump              D [range]
enter             E address [list]
fill              F range list
go                G [=address] [addresses]
hex               H value1 value2
input             I port
load              L [address] [drive] [firstsector] [number]
move              M range address
name              N [pathname] [arglist]
output            O port byte
proceed           P [=address] [number]
quit              Q
register          R [register]
search            S range list
trace             T [=address] [value]
unassemble        U [range]
write             W [address] [drive] [firstsector] [number]
allocate expanded memory                XA [#pages]
deallocate expanded memory              XD [handle]
map expanded memory pages               XM [Lpage] [Ppage]
                                           [handle]
display expanded memory status          XS
```

> **TIP:** See
>
> www.geocities.com/thestarman3/asm/debug/debug.htm
>
> or
>
> www.ping.be/~ping0751/debug.htm
>
> for more information on DEBUG.

For the use that DEBUG is put to here, to produce a hex dump, the only commands needed are D for Dump and Q for Quit.

To produce a hex dump of a file called TEST.DAT simply issue the command

```
C:\> DEBUG TEST.DAT
```

at the command prompt then use the command D (then enter). You can 'dump' from any address and any size, the command D 100 1000 dumps from address 100 and dumps 1000 bytes, both numbers in hex.

If you need the hex dump in a file, you should first prepare a command file that contains just the data below. Assume this command file is called CMD.DAT, it is made using the commands at the DOS prompt

```
C:\> COPY CON CMD.DAT
D
Q
CTRL Z
```

where CTRL Z means press the CTRL and Z keys together.

You can now produce a hex dump in a file by using I/O redirection as in the command:

```
C:\> DEBUG TEST.DAT < CMD.DAT > TESTHEXDUMP.TXT
```

The < character means take input from your command file called CMD.DAT and the > character means place or redirect the output into whatever filename you provide, in this case TESTHEXDUMP.TXT

If you need to dump more than 100 hex bytes, the file CMD.DAT will have to contain

```
D 100 200
Q
CTRL Z
```

where the 200 refers to 200 hex bytes in length. If you need to quote a length in the D command, the starting address (usually 100) must also appear.

The file TESTHEXDUMP.TXT now contains the hex dump of TEST.DAT

Appendix E Hex, binary, decimal and ASCII character set

E.1 ASCII character set

Before the days of computing, communication systems required each character to be sent as a code. Simple systems used 1s and 0s for transmission just like today so binary numbers were used to encode characters. You could not send an 'A' character directly but you could send binary 1000001 in its place. This eventually led to a 'standard' set of characters that were used to control printing devices before the widespread use of VDUs. ASCII stands for American Standard Code for Information Interchange but there are other character encoding systems around like EBCDIC and LICS and they work in a similar way but ASCII is the most widespread.

In ASCII, the codes from 0 to 31 were called 'Control Characters'. These were used to control the movement of the old mechanical printers so we have terms like 'Carriage Return' (now known as Enter or just Return) that actually caused the carriage that held the paper to return to the left-hand side. Understanding this historical basis of the control characters helps you to understand the names they are given which now seem a little odd. If a Control Character (written as CTRL A, etc.) is sent to a printer or screen, it usually results in an action rather than a printable character. Because some of the codes only have real meaning for mechanical printers, some modern uses do not always make sense of the original name.

Before the widespread use of Microsoft Windows, most machines responded directly to these control characters. As an experiment, try opening a DOS window and type a command. Instead of pressing the Enter key, press CTRL M instead, you should find it does the same thing as pressing Enter. The Enter key is just a CTRL M key in DOS. If you try this using Microsoft Word, CTRL M has a different effect. If you are using UNIX or Linux, try using CTRL H in place of the backspace key, it should work unless it has been remapped on your machine.

Characters in ASCII are easy to remember, they run from A = 65 to Z = 90. This may look like an odd choice of numbers until you convert the 65 into binary and get 1000001, i.e. 64 + 1. This means that any letter is easy to calculate, it is 64 plus the position in the alphabet. M is

Table E.1 The ASCII control characters

Dec	Hex	Key-board	Binary	Description	
0	0	CTRL @	00000	NUL	Null Character
1	1	CTRL A	00001	SOH	Start of Heading
2	2	CTRL B	00010	STX	Start of Text
3	3	CTRL C	00011	ETX	End of Text
4	4	CTRL D	00100	EOT	End of Transmission
5	5	CTRL E	00101	ENQ	Enquiry
6	6	CTRL F	00110	ACK	Acknowledge
7	7	CTRL G	00111	BEL	Bell or beep
8	8	CTRL H	01000	BS	Back Space
9	9	CTRL I	01001	HT	Horizontal Tab
10	A	CTRL J	01010	LF	Line Feed
11	B	CTRL K	01011	VT	Vertical Tab
12	C	CTRL L	01100	FF	Form Feed
13	D	CTRL M	01101	CR	Carriage Return
14	E	CTRL N	01110	SO	Shift Out
15	F	CTRL O	01111	SI	Shift In
16	10	CTRL P	10000	DLE	Date Link Escape
17	11	CTRL Q	10001	DC1	Device Control 1
18	12	CTRL R	10010	DC2	Device Control 2
19	13	CTRL S	10011	DC3	Device Control 3
20	14	CTRL T	10100	DC4	Device Control 4
21	15	CTRL U	10101	NAK	Negative Acknowledge
22	16	CTRL V	10110	SYN	Synchronous Idle
23	17	CTRL W	10111	ETB	End of Transmission Block
24	18	CTRL X	11000	CAN	Cancel
25	19	CTRL Y	11001	EM	End Medium
26	1A	CTRL Z	11010	SUB	Substitute or EOF End Of File
27	1B		11011	ESC	Escape
28	1C		11100	FS	File Separator
29	1D		11101	GS	Group Separator
30	1E		11110	RS	Record Separator
31	1F		11111	US	Unit Separator

the 13th letter in the alphabet so in ASCII, M = 64 + 13 = 77. To make it lower-case, just add 32. This is a good choice as 32 encodes as a single binary digit. Lower-case m is then 64 + 32 + 13 = 109. Of course it would be better to use hex, so A = 41, M = 4D, a = 61, m = 6D, etc. Numerals are just as easy, '0' encodes as 48, '1' encodes as 48 + 1 = 49, etc.

The full set of 7-bit printable ASCII characters is shown in Table E.2.

Table E.2 The full set of 7 bit printable ASCII characters is shown here

Char	Dec	Hex	Binary	Char	Dec	Hex	Binary	Char	Dec	Hex	Binary
Space	32	20	100000								
!	33	21	100001	A	65	41	1000001	a	97	61	1100001
''	34	22	100010	B	66	42	1000010	b	98	62	1100010
#	35	23	100011	C	67	43	1000011	c	99	63	1100011
$	36	24	100100	D	68	44	1000100	d	100	64	1100100
%	37	25	100101	E	69	45	1000101	e	101	65	1100101
&	38	26	100110	F	70	46	1000110	f	102	66	1100110
'	39	27	100111	G	71	47	1000111	g	103	67	1100111
(	40	28	101000	H	72	48	1001000	h	104	68	1101000
)	41	29	101001	I	73	49	1001001	I	105	69	1101001
*	42	2A	101010	J	74	4A	1001010	j	106	6A	1101010
+	43	2B	101011	K	75	4B	1001011	k	107	6B	1101011
,	44	2C	101100	L	76	4C	1001100	l	108	6C	1101100
-	45	2D	101101	M	77	4D	1001101	m	109	6D	1101101
.	46	2E	101110	N	78	4E	1001110	n	110	6E	1101110
/	47	2F	101111	O	79	4F	1001111	o	111	6F	1101111
0	48	30	110000	P	80	50	1010000	p	112	70	1110000
1	49	31	110001	Q	81	51	1010001	q	113	71	1110001
2	50	32	110010	R	82	52	1010010	r	114	72	1110010
3	51	33	110011	S	83	53	1010011	s	115	73	1110011
4	52	34	110100	T	84	54	1010100	t	116	74	1110100
5	53	35	110101	U	85	55	1010101	u	117	75	1110101
6	54	36	110110	V	86	56	1010110	v	118	76	1110110
7	55	37	110111	W	87	57	1010111	w	119	77	1110111
8	56	38	111000	X	88	58	1011000	x	120	78	1111000
9	57	39	111001	Y	89	59	1011001	y	121	79	1111001
:	58	3A	111010	Z	90	5A	1011010	z	122	7A	1111010
;	59	3B	111011	[	91	5B	1011011	{	123	7B	1111011
<	60	3C	111100	\	92	5C	1011100	I	124	7C	1111100
=	61	3D	111101	]	93	5D	1011101	}	125	7D	1111101
>	62	3E	111110	^	94	5E	1011110	~	126	7E	1111110
?	63	3F	111111	_	95	5F	1011111	del	127	7F	1111111
@	64	40	1000000	'	96	60	1100000				

You will notice that the codes only extend to 127. This is because the original ASCII only used 7 binary digits and was referred to as a 7-bit code. While there is some standardization of the codes 128 to 255, some machines will give different characters for codes 128 to 255, for instance older machines will give an é for code 130 while more modern machines will give an é for code 233.

While it is not important to remember ASCII codes, it is often useful, especially when writing text or string handling parts of programs. If you remember that 'A' = 64 + alphabet position (40 in hex) and that 'a' = 'A' + 32 ('A' + 20 in hex) you can work out all of the alphabet. The '0' character is 48 and the digits are 48 + their value. If you also remember that a Carriage Return is 13 (0D hex) and that Line Feed is

10 (0A hex) you will be able to remember about half of the codes and be able to interpret some hex-dumped files.

Although ASCII is a 7-bit code, Windows uses codes 128 to 255 for other characters as shown in Table E.3. This table was produced using the CHAR function in Excel 97.

E.2 Unicode

ASCII characters, although universally accepted, present one serious problem, there are not sufficient characters to cover all symbols and characters from different languages. The solution adopted until the introduction of *Unicode* was to set up each computer to have its own character set according to country or language. This makes it harder to communicate files between computers set up for different countries; try finding the pound sign on an American keyboard! Unicode uses 16-bit characters so there are $2^{16} = 65\,536$ possible characters, more than enough to cover all the world's main languages. The ASCII character set has been incorporated so character 65 is still an 'A' but the 65 is a 16-bit value. The Unicode standard is developing all the time, the latest situation is presented on their web page at www.unicode.org/unicode/standard/standard.html. This describes the current version 3 which defines 49 194 different characters and the work in progress to add more.

Conversion of ASCII to Unicode is very easy as the codes are simply changed 8 bit into 16 bit. Conversion from Unicode to ASCII may result in the loss of data as ASCII cannot support more than 256 different characters. Some operating systems will work with both character sets, the more modern ones will use Unicode as the native code.

Unicode is developing all the time with new characters being added, etc. See the latest information on the website www.unicode.org

Table E.3 Windows 8-bit character codes 128 to 255

Decimal	Hex	Binary	Character	Decimal	Hex	Binary	Character
128	80	10000000	€	192	C0	11000000	À
129	81	10000001	□	193	C1	11000001	Á
130	82	10000010	‚	194	C2	11000010	Â
131	83	10000011	ƒ	195	C3	11000011	Ã
132	84	10000100	„	196	C4	11000100	Ä
133	85	10000101	…	197	C5	11000101	Å
134	86	10000110	†	198	C6	11000110	Æ
135	87	10000111	‡	199	C7	11000111	Ç
136	88	10001000	ˆ	200	C8	11001000	È
137	89	10001001	‰	201	C9	11001001	É
138	8A	10001010	Š	202	CA	11001010	Ê
139	8B	10001011	<	203	CB	11001011	Ë
140	8C	10001100	Œ	204	CC	11001100	Ì
141	8D	10001101	□	205	CD	11001101	Í
142	8E	10001110	Ž	206	CE	11001110	Î
143	8F	10001111	□	207	CF	11001111	Ï
144	90	10010000	□	208	D0	11010000	Đ
145	91	10010001	'	209	D1	11010001	Ñ
146	92	10010010	'	210	D2	11010010	Ò
147	93	10010011	"	211	D3	11010011	Ó
148	94	10010100	"	212	D4	11010100	Ô
149	95	10010101	•	213	D5	11010101	Õ
150	96	10010110	–	214	D6	11010110	Ö
151	97	10010111	—	215	D7	11010111	×
152	98	10011000	˜	216	D8	11011000	Ø
153	99	10011001	™	217	D9	11011001	Ù
154	9A	10011010	š	218	DA	11011010	Ú
155	9B	10011011	>	219	DB	11011011	Û
156	9C	10011100	œ	220	DC	11011100	Ü
157	9D	10011101	□	221	DD	11011101	Ý
158	9E	10011110	ž	222	DE	11011110	Þ
159	9F	10011111	Ÿ	223	DF	11011111	β
160	A0	10100000		224	E0	11100000	à
161	A1	10100001	¡	225	E1	11100001	á
162	A2	10100010	¢	226	E2	11100010	â
163	A3	10100011	£	227	E3	11100011	ã
164	A4	10100100	¤	228	E4	11100100	ä
165	A5	10100101	¥	229	E5	11100101	å
166	A6	10100110	¦	230	E6	11100110	æ
167	A7	10100111	§	231	E7	11100111	ç
168	A8	10101000	¨	232	E8	11101000	è
169	A9	10101001	©	233	E9	11101001	é
170	AA	10101010	ª	234	EA	11101010	ê
171	AB	10101011	«	235	EB	11101011	ë
172	AC	10101100	¬	236	EC	11101100	ì
173	AD	10101101	-	237	ED	11101101	í
174	AE	10101110	®	238	EE	11101110	î
175	AF	10101111	¯	239	EF	11101111	ï
176	B0	10110000	°	240	F0	11110000	ð
177	B1	10110001	±	241	F1	11110001	ñ
178	B2	10110010	²	242	F2	11110010	ò
179	B3	10110011	³	243	F3	11110011	ó
180	B4	10110100	´	244	F4	11110100	ô
181	B5	10110101	µ	245	F5	11110101	õ
182	B6	10110110	¶	246	F6	11110110	ö
183	B7	10110111	·	247	F7	11110111	÷
184	B8	10111000	¸	248	F8	11111000	ø
185	B9	10111001	¹	249	F9	11111001	ù
186	BA	10111010	º	250	FA	11111010	ú
187	BB	10111011	»	251	FB	11111011	û
188	BC	10111100	¼	252	FC	11111100	ü
189	BD	10111101	½	253	FD	11111101	ý
190	BE	10111110	¾	254	FE	11111110	þ
191	BF	10111111	¿	255	FF	11111111	ÿ

Appendix F IBM POST and diagnostic error codes

F.1 Audible BIOS error codes

F.1.1 Original IBM BIOS

Beeps	Fault indicated
One short beep	Normal POST – no error
Two short beeps	POST error – see screen for error code
No beeps	Power missing, loose card or short circuit
Continuous beep	Power missing, loose card or short circuit
Repeating short beep	Power missing, loose card or short circuit
One long and one short beep	System board error
One long and two short beeps	Video (mono/CGA display adapter)
One long and three short beeps	Video (EGA display adapter)
Three long beeps	Keyboard error
One beep	Blank/incorrect display, video display circuitry

F.1.2 Phoenix BIOS

Beeps	Fault indicated
One, one and three beeps	CMOS read/write failure
One, one and four beeps	ROM BIOS checksum failure
One, two and one beep	Programmable interval timer failure
One, two and two beeps	DMA initialization failure
One, two and three beeps	DMA page register read/write failure
One, three and one beep	RAM refresh verification error
One, three and three beeps	First 64k RAM chip/data line failure
One, three and four beeps	First 64k odd/even logic failure
One, four and one beep	Address line failure first 64k RAM
One, four and two beeps	Parity failure first 64k RAM
One, four and three beeps	Fail-safe timer feature (EISA only)
One, four and four beeps	Software NMI port failure (EISA only)
Two, one and up to four beeps	First 64k RAM chip/data line failure (bits 0 to 3 respectively)
Two, two and up to four beeps	First 64k RAM chip/data line failure (bits 4 to 7 respectively)

Two, three and up to four beeps	First 64k RAM chip/data line failure (bits 8 to 11 respectively)
Two, four and up to four beeps	First 64k RAM chip/data line failure (bits 12 to 15 respectively)
Three, one and one beep	Slave DMA register failure
Three, one and two beeps	Master DMA register failure
Three, one and three beeps	Master interrupt mask register failure
Three, one and four beeps	Slave interrupt register failure
Three, two and four beeps	Keyboard controller test failure
Three, three and four beeps	Screen initialization failure
Three, four and one beep	Screen retrace test failure
Four, two and one beep	Timer tick failure
Four, two and two beeps	Shutdown test failure
Four, two and three beeps	Gate A20 failure
Four, two and four beeps	Unexpected interrupt in protected mode
Four, three and one beep	RAM text address failure
Four, three and three beeps	Interval timer channel 2 failure
Four, three and four beeps	Time of day clock failure
Four, four and three beeps	Maths coprocessor failure

F.1.3 American Megatrends' AMI BIOS

Beeps	Fault indicated
1	The memory refresh circuitry has failed
2	Parity errors have been detected in the first 64 KB of memory
3	A failure has occurred within the first 64 KB of memory
4	System timer failure: timer 1 on the mainboard does not work properly
5	The CPU has generated an undetectable error
6	8042 Gate-A20 failure: BIOS cannot switch the CPU into protected mode
7	The CPU has generated an exception error
8	The video adapter is missing, or the memory on the adapter has generated a failure
9	The ROM checksum value does not match the value in BIOS
10	The shutdown register for CMOS interrupt channel 2 has failed POST; the system board cannot retrieve CMOS contents during POST
11	Level-2 cache memory has failed the tests, and has been disabled

2 short	POST has failed, caused by a failure of one of the hardware tests
1 long 2 short	Failure in video system: a checksum error was encountered in video BIOS ROM, or a horizontal retrace failure has been encountered
1 long 3 short	Failure in video system: the video DAC, the monitor detection procedure or the video RAM has failed
1 long	POST procedures have passed

F.1.4 AST Research BIOS

Beeps	Fault indicated
1 short	Low level processor verification test failed (POST 1)
2 short	Clearing keyboard controller buffers failed (POST 2)
3 short	Keyboard controller reset failed (POST 3)
4 short	Low level keyboard controller interface test (POST 4)
5 short	Reading data from keyboard controller failed (POST 5)
6 short	System board support chip initialization failed (POST 6)
7 short	Processor register read/write verify test failed (POST 7)
8 short	CMOS timer initialization failed (POST 8)
9 short	ROM BIOS checksum test failed (POST 9)
10 short	Initialize primary video (POST 10)
11 short	8254 timer channel 0 test failed (POST 11)
12 short	8254 timer channel 1 test failed (POST 12)
13 short	8254 timer channel 2 test failed (POST 13)
14 short	CMOS power-on and time test failed (POST 14)
15 short	CMOS shutdown byte test failed (POST 15)
1 long	DMA channel 0 test failed (POST 16)
1 long 1 short	DMA channel 1 test failed (POST 17)
1 long 2 short	DMA page register test failed (POST 18)
1 long 3 short	Keyboard controller interface test failed (POST 19)
1 long 4 short	Memory refresh toggle test failed (POST 20)
1 long 5 short	First 64 KB memory test failed (POST 21)
1 long 6 short	Setup interrupt vector table failed (POST 22)
1 long 7 short	Video initialization failed (POST 23)
1 long 8 short	Video memory test failed (POST 24)

F.1.5 AST Enhanced BIOS

Short	Long	Short	Fault indicated
3	1	–	Flash loader failure
3	2	–	Failure in system board component
3	3	–	Failure in system board component
3	4	–	Memory failure
3	5	–	Video failure
0	6	–	Flash BIOS update error
–	2	x	AST low level diagnostics

F.1.6 AST Phoenix BIOS

Beeps	Fault indicated
1-1-3	CMOS read/write failure
1-1-4	ROM BIOS checksum failure
1-2-1	Programmable interval timer failure
1-2-2	DMA initialization failure
1-2-3	DMA page register read/write failure
1-3-1	RAM refresh verification failure
1-3-3	First 64 KB RAM chip or data or data line failure
1-3-4	First 64 KB RAM odd/even logic failure
1-4-1	First 64 KB RAM address line failure
1-4-2	First 64 KB RAM parity failure
2-1-1	First 64 KB RAM failure bit 0
2-1-2	First 64 KB RAM failure bit 1
2-1-3	First 64 KB RAM failure bit 2
2-1-4	First 64 KB RAM failure bit 3
2-2-1	First 64 KB RAM failure bit 4
2-2-2	First 64 KB RAM failure bit 5
2-2-3	First 64 KB RAM failure bit 6
2-2-4	First 64 KB RAM failure bit 7
2-3-1	First 64 KB RAM failure bit 8
2-3-2	First 64 KB RAM failure bit 9
2-3-3	First 64 KB RAM failure bit A
2-3-4	First 64 KB RAM failure bit B
2-4-1	First 64 KB RAM failure bit C
2-4-2	First 64 KB RAM failure bit D
2-4-3	First 64 KB RAM failure bit E
2-4-4	First 64 KB RAM failure bit F
3-1-1	Slave DMA register failure
3-1-2	Master DMA register failure
3-1-3	Slave interrupt mask register failure
3-1-4	Slave interrupt mask failure
3-2-4	Keyboard controller test failure

3-3-4	Screen memory test failure
3-4-1	Screen initialization failure
3-4-2	Screen retrace test failure
3-4-3	Failure searching for video ROM
4-2-1	No timer tick
4-2-3	Gate-A20 failure
4-2-4	Unexpected interrupt in protected mode

Appendix G Reference section

G.1 Nanoseconds and other named fractions

G.1.1 Powers of 10 and their names

Factor of 10	Value	Prefix	Symbol
10^{-18}	0.00000 00000 00000 001	atto	a
10^{-15}	0.00000 00000 00001	femto	f
10^{-12}	0.00000 00000 01	pico	p
10^{-9}	0.00000 0001	nano	n
10^{-6}	0.00000 1	micro	μ
10^{-3}	0.001	milli	m
10^{-2}	0.01	centi	c
10^{-1}	0.1	deci	d
10	10	deca	da
10^{2}	100	hecto	h
10^{3}	1000	kilo	k
10^{6}	10000 00	mega	M
10^{9}	10000 00000	giga	G
10^{12}	10000 00000 000	tera	T
10^{15}	10000 00000 00000 0	peta	P
10^{18}	10000 00000 00000 0000	exa	E
10^{21}	10000 00000 00000 00000 00	zetta	Z
10^{24}	10000 00000 00000 00000 00000	yotta	Y

G.1.2 Powers of 2 and their names

Power of 2	Number of bytes	Symbol	Name	Example
2^{10}	1 024	kb	kilo	kilobytes
2^{20}	1 048 576	Mb	mega	megabytes
2^{30}	1 073 741 824	Gb	giga	gigabytes
2^{40}	1 099 511 627 776	Tb	tera	terabytes
2^{50}	1 125 899 906 843 624	Pb	peta	petabytes
2^{60}	1 152 921 504 607 870 976	Eb	exa	exabytes
2^{70}	1 180 591 620 718 458 879 424	Zb	zetta	zettabytes
2^{80}	1 208 925 819 615 701 892 530 176	Yb	yotta	yottabytes

G.2 Glossary of terms

Accelerator A board which replaces the CPU with circuitry to increase the speed of processing.

Access time The time taken to retrieve data from a memory/storage device, i.e. the elapsed time between the receipt of a read signal at the device and the placement of valid data on the bus. Typical access times for semiconductor memory devices are in the region 100 ns to 200 ns while average access times for magnetic disks typically range from 10 ms to 50 ms.

Accumulator A register within the central processing unit (CPU) in which the result of an operation is placed.

Acknowledge (ACK) A signal used in serial data communications which indicates that data has been received without error.

Active high A term used to describe a signal which is asserted in the high (logic 1) state.

Active low A term used to describe a signal which is asserted in the low (logic 0) state.

Address A reference to the location of data in memory or within I/O space. The CPU places addresses (in binary coded form) on the address bus.

Address bus The set of lines used to convey address information. The IBM-PC bus has 20 address lines (A0 to A19) and these are capable of addressing more than a million address locations. One byte of data may be stored at each address.

Address decoder A hardware device (often a single integrated circuit) which provides chip select or chip enable signals from address patterns which appear on an address bus.

Address selection The process of selecting a specific address (or range of addresses). In order to prevent conflicts, expansion cards must usually be configured (by means of DIP switches or links) to unique addresses within the I/O address map.

AGP Advanced graphics port. Used in addition to PCI to communicate with the CPU and RAM.

Amplifier A circuit or device which increases the power of an electrical signal.

Analogue The representation of information in the form of a continuously variable quantity (e.g. voltage).

AND Logical function which is asserted (true) when all inputs are simultaneously asserted.

ANSI character set The American National Standard Institute's character set which is based on an 8-bit binary code and which provides 256 individual characters. See also ASCII.

Archive A device or medium used for storage of data which need not be instantly accessible (e.g. a tape cartridge).
ASCII A code which is almost universally employed for exchanging data between microcomputers. Standard ASCII is based on a 7-bit binary code and caters for alphanumeric characters (both upper and lower case), punctuation and special control characters. Extended ASCII employs an eighth bit to provide an additional 128 characters (often used to represent graphic symbols).
Assembly language A low-level programming language which is based on mnemonic instructions. Assembly language is often unique to a particular microprocessor or microprocessor family.
Asserted A term used to describe a signal when it is in its logically true state (i.e. logic 1 in the case of an active high signal or logic 0 in the case of an active low signal).
Asynchronous transmission A data transmission method in which the time between transmitted characters is arbitrary. Transmission is controlled by start and stop bits (no additional synchronizing or timing information is required).
ATAPI The ATAPI (or Advanced Technology Attachment Packet Interface) standard provides a simple means of connecting a CD-ROM drive to an EIDE adapter. Without such an interface, a CD-ROM drive will require either a dedicated interface card or an interface provided on a sound card.
AUTOEXEC.BAT A file which contains a set of DOS commands and/or program names which is executed automatically whenever the system is initialized and provides a means of configuring a system.
Backup A file or disk copy made in order to avoid the accidental loss, damage, or erasure of programs and/or data.
Basic input output system (BIOS) The BIOS is the part of the operating system which handles communications between the microcomputer and peripheral devices (such as keyboard, serial port, etc.). The BIOS is supplied as firmware and is contained in a read-only memory (ROM).
Batch file A file containing a series of DOS commands which are executed when the filename is entered after the DOS prompt. Batch files are given a BAT file extension. A special type of batch file (AUTOEXEC.BAT) is executed (when present) whenever a system is initialized. See also AUTOEXEC.BAT.
Baud rate The speed at which serial data is transferred between devices.
Binary file A file which contains binary data (i.e. a direct memory image). This type of file is used for machine readable code, program overlays and graphics screens.
Bit A contraction of 'binary digit'; a single digit in a binary number.

Boot The name given to the process of loading and initializing an operating system (part of the operating system is held on disk and must be loaded from disk into RAM on power-up).

Boot record A single-sector record present on a disk which conveys information about the disk and instructs the computer to load the requisite operating system files into RAM (thus booting the machine).

Buffer In a hardware context, a buffer is a device which provides a degree of electrical isolation at an interface. The input to a buffer usually exhibits a much higher impedance than its output (see also Driver). In a software context, a buffer is a reserved area of memory which provides temporary data storage and thus may be used to compensate for a difference in the rate of data flow or time of occurrence of events.

Bus An electrical highway for signals which have some common function. Most microprocessor systems have three distinct buses: an address bus, data bus and control bus. A local bus can be used for high-speed data transfer between certain devices (e.g. CPU, graphics processors and video memory).

Byte A group of 8 bits which are operated on as a unit.

Cache A high-speed random-access memory which is used to store copies of the data from the most recent main memory or hard disk accesses. Subsequent accesses fetch data from this area rather than from the slower main memory or hard disk.

Central processing unit (CPU) The part of a computer that decodes instructions and controls the other hardware elements of the system. The CPU comprises a control unit, arithmetic/logic unit and internal storage. In microcomputers, a microprocessor acts as the CPU. See also Microprocessor.

Channel A path along which signals or data can be sent.

Character set The complete range of characters (letters, numbers and punctuation) which are provided within a system. See also ANSI and ASCII.

Checksum Additional binary digits appended to a block of data. The value of the appended digits is derived from the sum of the data present within the block. This technique provides a means of error checking (validation).

Chip The term commonly used to describe an integrated circuit.

CISC The term CISC refers to a 'complex instruction set computer' – the standard Intel family of CPUs all conform to this model rather than the alternative 'reduced instruction set computer' (RISC). There is much debate about the pros and cons of these two design methodologies but, in fact, neither of these two contrasting approaches has actually demonstrated clear superiority over the other! See also 'CISC'.

Clock A source of timing signals used for synchronizing data transfers within a microprocessor or microcomputer system.

Cluster A unit of space allocated on the surface of a disk. The number of sectors which make up a cluster varies according to the DOS version and disk type. See also Sector.
Command An instruction (entered from the keyboard or contained within a batch file) which will be recognized and executed by a system. See also Batch file.
Common A return path for a signal (often ground).
CONFIG.SYS A file which contains DOS configuration commands which are used to configure the system at startup. The CONFIG.SYS file specifies device drivers which are loaded during initialization and which extend the functionality of a system by allowing it to communicate with additional items of hardware. See also Device driver.
Controller A sub-system within a microcomputer which controls the flow of data between the system and an I/O or storage device (e.g. a CRT controller, hard disk controller, etc.). A controller will generally be based on one, or more, programmable VLSI devices.
Coprocessor A second processor which shares the same instruction stream as the main processor. The coprocessor handles specific tasks (e.g. mathematics) which would otherwise be performed less efficiently (or not at all) by the main processor.
Cylinder The group of tracks which can be read from a hard disk at any instant of time (i.e. without steeping the head in or out). In the case of a floppy disk (where there are only two surfaces), each cylinder comprises two tracks. In the case of a typical IDE hard disk, there may be two platters (i.e. four surfaces) and thus four tracks will be present within each cylinder.
Daisy chain A method of connection in which signals move in a chained fashion from one device to another. This form of connection is commonly used with disk drives.
Data A general term used to describe numbers, letters and symbols present with a computer system. All such information is ultimately represented by patterns of binary digits.
Data bus A highway (in the form of multiple electrical conductors) which conveys data between the different elements within a microprocessor system.
Data file A file which contains data (rather than a program) and which is used by applications such as spreadsheet and database applications. Note that data may or may not be stored in directly readable ASCII form.
Device A hardware component such as a memory card, sound card, modem, or graphics adapter.
Device driver A term used to describe memory resident software (specified in the CONFIG.SYS system file) which provides a means of

interfacing specialized hardware (e.g. expanded memory adapters). See CONFIG.SYS.

Direct memory access A method of fast data transfer in which data moves between a peripheral device (e.g. a hard disk) and main memory without direct control of the CPU.

Directory A catalogue of disk files (containing such information as filename, size, attributes and date/time of creation). The directory is stored on the disk and updated whenever a file is amended, created, or deleted. A directory entry usually comprises 32 bytes for each file.

DIP switch A miniature PCB mounted switch that allows configuration options (such as IRQ or DMA settings) to be selected.

Disk operating system (DOS) A group of programs which provide a low-level interface with the system hardware (particularly disk I/O). Routines contained within system resident portions of the operating system may be used by the programmer. Other programs provided as part of the system include those used for formatting disks, copying files, etc.

Double word A data value which comprises a group of 32 bits (or two words). See also 'Word'.

DRAM DRAM (or dynamic random access memory) refers to the semiconductor read/write memory of a PC. DRAM requires periodic 'refreshing' and therefore tends not to offer the highest speeds required of specialized memories (such as cache memory). DRAM is, however, relatively inexpensive.

Driver In a software context, a driver is a software routine which provides a means of interfacing a specialized hardware device (see also Device driver). In a hardware context, a driver is an electrical circuit which provides an electrical interface between an output port and an output transducer. A driver invariably provides power gain (i.e. current gain and/or voltage gain). See also Amplifier.

EIDE EIDE (or Enhanced Integrated Drive Electronics) is the most widely used interface for connecting hard disk drives to a PC. Most motherboards now incorporate an on-board EIDE controller rather than having to make use of an adapter card. This allows one or two hard disk drives to be connected directly to the motherboard.

Expanded memory (EMS memory) Memory which is additional to the conventional 'base' memory available within the system. This memory is 'paged' into the base memory space whenever it is accessed. The EMS specification uses four contiguous 16K pages of physical memory (64K total) to access up to 32M of expanded memory space. See also Expanded memory manager.

Expanded memory manager An expanded memory manager (such as EMM386.EXE included with MS-DOS 5.0 and later) provides a means of establishing and controlling the use of expanded memory (i.e.

memory above the DOS 1 Mbyte limit). Unlike DOS and Windows 3.1, Windows 95 incorporates its own memory management and thus EMM386 (or its equivalent) is not required. See also Expanded memory.

Extended memory (XMS memory) Memory beyond the 1M byte range ordinarily recognized by MS-DOS. The XMS memory specification resulted from collaboration between Lotus, Intel and Microsoft (sometimes known as LIM specification).

File Information (which may comprise ASCII encoded text, binary coded data and executable programs) stored on a floppy or hard disk. Files may be redirected from one logical device to another using appropriate DOS commands.

File allocation table (FAT) The file allocation table (or FAT) provides a means of keeping track of the physical location of files stored on a floppy disk or hard disk. Part of the function of DOS is to keep the FAT up to date whenever a file operation is carried out. DOS does not necessarily store files in physically contiguous clusters on a disk and it is the FAT that maintains the addresses of clusters occupied by a particular file. These clusters may, in fact, be scattered all over the surface of the disk (in which case we describe the file as having been 'fragmented').

File attributes Information which indicates the status of a file (e.g. hidden, read only, system, etc.).

Filter In a software context, a filter is a software routine which removes or modifies certain data items (or data items within a defined range). In a hardware context, a filter is an electrical circuit which modifies the frequency distribution of a signal. Filters are often categorized as low-pass, high-pass, band-pass, or band-stop depending upon the shape of their frequency response characteristic.

Firmware A program (software) stored in read-only memory (ROM). Firmware provides non-volatile storage of programs.

Fixed disk A disk which cannot be removed from its housing. Note that, while the terms 'hard' and 'fixed' are often used interchangeably, some forms of hard disk are exchangeable.

Font A set of characters (letters, numbers and punctuation) with a particular style and size.

Format The process in which a magnetic disk is initialized so that it can accept data. The process involves writing a magnetic pattern of tracks and sectors to a blank (uninitialized) disk. A disk containing data can be reformatted, in which case all data stored on the disk will be lost. An MS-DOS utility program (FORMAT.COM) is supplied in order to carry out the formatting of floppy disks (a similar utility is usually provided for formatting the hard disk).

Graphics adapter An option card which provides a specific graphics capability (e.g. CGA, EGA, HGA, VGA). Graphics signal generation is not normally part of the functionality provided within a system mother board.

Handshake An interlocked sequence of signals between peripheral devices in which a device waits for an acknowledgement of the receipt of data before sending new data.

Hard disk A non-flexible disk used for the magnetic storage of data and programs. See also Fixed disk.

Hardware The physical components (e.g. system board, keyboard, etc.) which make up a microcomputer system.

High state The more positive of the two voltage levels used to represent binary logic states. A high state (logic 1) is generally represented by a voltage in the range 2.0 V to 5.0 V.

High memory The first 64K of extended memory. This area is used by some DOS applications and also by Windows. See Extended memory.

IDE IDE (or Integrated Drive Electronics) is the forerunner of the EIDE interface used in most modern PCs. See EIDE.

Input/output (I/O) Devices and lines used to transfer information to and from external (peripheral) devices.

Integrated circuit An electronic circuit fabricated on a single wafer (chip) and packaged as a single component.

Interface A shared boundary between two or more systems, or between two or more elements within a system. In order to facilitate interconnection of systems, various interface standards are adopted (e.g. RS-232 in the case of asynchronous data communications).

Interleave A system of numbering the sectors on a disk in a non-consecutive fashion in order to optimize data access times.

Interrupt A signal generated by a peripheral device when it wishes to gain the attention of the CPU. The Intel 80x86 family of microprocessors support both software and hardware interrupts. The former provide a means of invoking BIOS and DOS services while the latter are generally managed by an interrupt controller chip (e.g. 8259).

ISA ISA (or Industry Standard Architecture) is the long-surviving standard for connecting multiple interface adapters to the PC bus. Due to speed limitations, the ISA bus is no longer used for hardware that requires fast data throughput and local bus schemes (such as VL bus or PCI bus) are much preferred.

Joystick A device used for positioning a cursor, pointer, or output device using switches or potentiometers which respond to displacement of the stick in the X and Y directions.

Jumper Jumpers, like DIP switches, provide a means of selecting configuration options on adapter cards. See DIP switch.

Keyboard buffer A small area in memory which provides temporary storage for keystrokes. See Buffer.

Kilobyte (K) 1024 bytes (note that $2^{10} = 1024$).

Logical device A device which is normally associated with microcomputer I/O, such as the console (which comprises keyboard and display) and printer.
Low state The more negative of the two voltage levels used to represent the binary logic states. A low state (logic 0) is generally represented by a voltage in the range 0 V to 0.8 V.
Megabyte (M) 1 048 576 bytes (note that $2^{20} = 1\,048\,576$). The basic addressing range of the 8086 (which has 20 address bus lines) is 1 Mbyte.
Memory That part of a microcomputer system into which information can be placed and later retrieved. Storage and memory are interchangeable terms. Memory can take various forms including semiconductor (RAM and ROM), magnetic (floppy and hard disks) and optical disks. Note that memory may also be categorized as read only (in which case data cannot subsequently be written to the memory) or read/write (in which case data can both be read from and written to the memory).
Memory resident program See TSR.
Microprocessor A central processing unit fabricated on a single chip.
MIDI The MIDI (or musical instrument digital interface) is the current industry standard for connecting musical instruments to a PC.
Modem A contraction of modulator–demodulator; a communications interface device that enables a serial port to be interfaced to a conventional voice-frequency telephone line.
Modified frequency modulation (MFM) A method of data encoding employed with hard disk storage. This method of data storage is 'self-clocking'.
Motherboard The motherboard (or system board) is the mother printed circuit board which provides the basic functionality of the microcomputer system including CPU, RAM and ROM. The system board is fitted with connectors which permit the installation of one, or more, option cards (e.g. graphics adapters, disk controllers, etc.).
Multimedia A combination of various media technologies including sound, video, graphics and animation.
Multitasking A process in which several programs are running simultaneously.
NAND Inverse of the logical AND function.
Negative acknowledge (NAK) A signal used in serial data communications which indicates that erroneous data has been received.
Network A system which allows two or more computers to be linked via a physical communications medium (e.g. coaxial cable) in order to exchange information and share resources.
Nibble A group of 4 bits which make up one half of a byte. A hexadecimal character can be represented by such a group.
Noise Any unwanted signal component which may appear superimposed on a wanted signal.

NOR Inverse of the logical OR function.

Operating system A control program which provides a low-level interface with the system hardware. The operating system thus frees the programmer from the need to produce hardware specific I/O routines (e.g. those associated with disk filing). See also Disk operating system.

Option card A printed circuit board (adapter card) which complies with the physical and electrical specification for a particular system and which provides the system with additional functionality (e.g. asynchronous communications facilities).

OR Logical function which is asserted (true) when any one or more of its inputs are asserted.

Page A contiguous area of memory of defined size (often 256 bytes but can be larger). See Expanded memory.

Paragraph Sixteen consecutive bytes of data. The segment address can be incremented to point to consecutive paragraphs of data.

Parallel interface (parallel port) A communications interface in which data is transferred a byte at a time between a computer and a peripheral device, such as a printer.

PCI The PCI (or peripheral component interconnect) standard provides a means of connecting 32-bit or 64-bit expansion cards to a motherboard. PCI expansion slots are available in most modern PCs.

PCMCIA The PCMCIA (or simply 'PC Card') standard provides a means of connecting a sub-miniature expansion card (such as a memory card or modem) to a laptop or book computer.

Peripheral An external hardware device whose activity is under the control of the microcomputer system.

Port A general term used to describe an interface circuit which facilitates transfer of data to and from external devices (peripherals).

Program A sequence of executable microcomputer instructions which have a defined function. Such instructions are stored in program files having EXE or COM extensions.

Propagation delay The time taken for a signal to travel from one point to another. In the case of logic elements, propagation delay is the time interval between the appearance of a logic state transition at the input of a gate and its subsequent appearance at the output.

Protocol A set of rules and formats necessary for the effective exchange of data between intelligent devices.

Random access An access method in which each word can be retrieved in the same amount of time (i.e. the storage locations can be accessed in any desired order). This method should be compared with sequential access in which access times are dependent upon the position of the data within the memory.

Random access memory (RAM) A term which usually refers to semiconductor read/write memory (in which access time is independent of actual storage address). Note that semiconductor read-only memory (ROM) devices also provide random access.

Read The process of transferring data to a processor from memory or I/O.

Read-only memory (ROM) A memory device which is permanently programmed. Erasable-programmable read only memory (EPROM) devices are popular for storage of programs and data in stand-alone applications and can be erased under ultraviolet light to permit reprogramming.

Register A storage area within a CPU, controller, or other programmable device, in which data (or addresses) are placed during processing. Registers will commonly hold 8-, 16- or 32-bit values.

RISC The term RISC refers to a 'reduced instruction set computer' – a computer based on a processor that accepts only a limited number of basic instructions but which decodes and executes them faster than the alternative technology (CISC). See also CISC.

Run length limited (RLL) A method of data encoding employed with hard disk storage. This method is more efficient than conventional MFM encoding.

Root directory The principal directory of a disk (either hard or floppy) which is created when the disk is first formatted. The root directory may contain the details of further sub-directories which may themselves contain yet more sub-directories, and so on.

SCSI The SCSI (or 'small computer systems interface') provides a means of interfacing up to eight peripheral devices (such as hard disks, CD-ROM drives and scanners) to a microcomputer system. With its roots in larger minicomputer systems, SCSI tends to be more complex and expensive in comparison with EIDE.

Sector The name given to a section of the circular track placed (during formatting) on a magnetic disk. Tracks are commonly divided into ten sectors. See also Format.

Segment 64 K bytes of contiguous data within memory. The starting address of such a block of memory may be contained within one of the four segment registers (DS, CS, SS, or ES).

Serial interface (serial port) A communications interface in which data is transferred a bit at a time between a computer and a peripheral device, such as a modem. In serial data transfer, a byte of data (i.e. 8 bits) is transmitted by sending a stream of bits, one after another. Furthermore, when such data is transmitted asynchronously (i.e. without a clock), additional bits must be added for synchronization together with further bits for error (parity) checking (if enabled).

Server A computer which provides network accessible services (e.g. hard disk storage, printing, etc).

Shell The name given to an item of software which provides the principal user interface to a system. The DOS program COMMAND.COM provides a simple DOS shell; however, later versions of MS-DOS and DR-DOS provide much improved graphical shells (DOSSHELL and VIEWMAX respectively).
Signal The information conveyed by an electrical quantity.
Signal level The relative magnitude of a signal when considered in relation to an arbitrary reference (usually expressed in volts, V).
SIMM SIMMs (or single in-line memory modules) are used to house the DRAM chips used in all modern PCs. The modular packaging and standard pin connections makes memory expansion very straightforward.
Software A series of computer instructions (i.e. a program).
Sub-directory A directory which contains details of a group of files and which is itself contained within another directory (or within the root directory).
System board See Motherboard.
Swap file A swap file is a file that resides on a hard disk and is used to provide 'virtual memory'. Swap files may be either 'permanent' or 'temporary'. See also Virtual memory.
System file A file that contains information required by DOS. Such a file is not normally shown in a directory listing.
Terminal emulation The ability of a microcomputer to emulate a hardware terminal.
TSR A terminate-and-stay-resident program (i.e. a program which, once loaded, remains resident in memory and which is available for execution from within another application).
UART UART (or universal asynchronous transmitter/receiver) is the name given to the chip that controls the PC's serial interface. Most modern PCs are fitted with 16 550 or 16 650 UARTs.
Upper memory The 384K region of memory which extends beyond the 640K of conventional memory. This region of memory is not normally available to applications and is reserved for system functions such as the video display memory. Some applications (such as Windows running in enhanced mode) can access unused portions of the upper memory area).
USB Universal serial bus.
Validation A process in which input data is checked in order to identify incorrect items. Validation can take several forms including range, character and format checks.
Verification A process in which stored data is checked (by subsequent reading) to see whether it is correct.
Virtual memory A technique of memory management which uses disk swap files to emulate random-access memory. The extent of RAM can

be increased by this technique by an amount which is equivalent to the total size of the swap files on the hard disk.

Visual display unit (VDU) An output device (usually based on a cathode ray tube) on which text and/or graphics can be displayed. A VDU is normally fitted with an integral keyboard in which case it is sometimes referred to as a console.

Volume label A disk name (comprising up to 11 characters). Note that hard disks may be partitioned into several volumes, each associated with its own logical drive specifier (i.e. C:, D:, E:, etc.).

VRAM VRAM (or video random access memory) is a high-speed type of DRAM fitted to a graphics controller card. This type of memory is preferred for the fast throughput of data which is essential when manipulating high-resolution screen images. See also DRAM.

Word A data value which comprises a group of 16 bits and which constitutes the fundamental size of data which an 8086 processor can accept and manipulate as a unit.

Write The process of transferring data from a CPU to memory or to an I/O device.

Appendix H Useful websites

Company		URL
Dr Solomon's Software	Anti-virus software, now part of McAfee	www.drsolomon.com
Hewlett Packard	Computer and peripherals manufacturer	www.hp.com
IBM	Computer and peripherals manufacturer	www.ibm.com
Symantec	Computer security and utility software supplier	www.symantec.com
McAfee	Computer security software supplier	www.mcafee.com
Carrera	Computer supplier	www.carrera.co.uk
Dan Technology	Computer supplier	www.dan.co.uk
Dell	Computer supplier	www.dell.com/uk
Gateway	Computer supplier	www.gw2k.co.uk
Kingston Technology	Computer supplier	www.kingston.com
Mesh	Computer supplier	www.meshplc.co.uk
Caldera	Domain names	www.caldera.co.uk
NEC	Electronics manufacturer	www.nec.com
Matrox	Graphics and networking manufacturer	www.matrox.com
Maxtor	Hard drive manufacturer	www.maxtor.com
Seagate	Hard drive manufacturer	www.seagate.com
Western Digital	Hard drive manufacturer	www.wdc.com
Intel	Microprocessor manufacturer	www.intel.com
AMD	Microprocessors	www.amd.com
Taxan	Monitor manufacturer	www.taxan.co.uk
Epson	Peripherals manufacturer	www.epson.co.uk
Microsoft	Software supplier	www.microsoft.com
Computer Information Centre	Sources of Computer Information	www.compinfo.co.uk

H.1 Search engines

Search Engine Information	www.searchenginewatch.com
Google	www.google.com
AllTheWeb.com (FAST)	www.alltheweb.com
Yahoo	www.yahoo.com
MSN Search	search.msn.com
Lycos	www.lycos.com
Ask Jeeves	www.askjeeves.com
AOL Search	search.aol.com
Teoma	www.teoma.com
WiseNut	www.wisenut.com

NB: Many companies maintain UK websites as well as sites in the USA. Where UK sites are known to exist these have been quoted.

Appendix I Processor types, sockets and families

The first few generations of Intel processor used '8' as the series name, so we get chips called the 8088, 8086, 80186, 80286, 80386 and 80486. Intel could not use the next number, 80586, as companies are not able to use numbers as trade marks, so they chose to use the name Pentium, pent referring to 5 as in pentathlete. This is unfortunate as the next combining form after pent is 'sex' for 6. Subsequent chips are called the Pentium II, then Pentium III, etc.

Intel 8086 (1978)
A true 16-bit processor with 20 address lines that could address up to 1 MB of RAM. The chip was available in 5, 6, 8 and 10 MHz versions.

Intel 8088 (1979)
The 8088 is almost identical to the 8086 except that it handles its address lines differently from the 8086. This chip was in the first IBM PC, and could work with the 8087 math coprocessor chip. These chips allowed real arithmetic, i.e. arithmetic using fractions.

NEC V20 and V30 (1981)
Clones of the 8088 and 8086.

Intel 80186 (1980)
A development of the 8086, a 16-bit version was available but never used in PCs. It was common on dedicated controller and embedded systems.

Intel 80286 (1982)
A 16-bit, 134 000 transistor chip that could address up to 16 MB of RAM. The 286 was the first processor that could use 'protected mode'. This allowed multitasking, the ability to run several processes more or less at the same time. The common operating system of the day, DOS or disc operating system did not use this feature but others did. The chip was used in huge numbers of AT (Advanced Technology) versions of the PC. It ran at 8, 10, 12.5 and later 20 MHz.

Intel 80386, usually known as the '386' (1985–1990)
This is a 32-bit processor containing 275 000 transistors and available in 16, 20, 25 and 33 MHz versions. The 32-bit address bus could address 4 GB of RAM and introduced 'pipelining' which allows the next instruction to be fetched from RAM while still working on the last one. In 1988, the 386SX, which was a simpler version of the 386, used the 16-bit data bus rather than the 32 bit but was slower. The advantage was that it used less power. The 386 still could not do real or floating point arithmetic; that required the help of an 80387 math coprocessor.

Intel 486 (1989–1994)
The 486 is a 32-bit processor using 1.2 million transistors and runs at twice the speed of a 386. It had an integrated math coprocessor now called a 'floating point unit' (FPU). Later a cheaper 486SX version was available with the FPU and a 486DX with the FPU. The 486 also contained an integrated 8 KB 'cache' that allows local storage of a few instructions. This means execution is faster as the processor does not have to fetch each instruction from (slow) RAM. In 1992, the i486DX2/50 and i486DX2/66 were released. The extra '2' indicates that the clock speed of the processor was doubled. The 486SL was also available, almost identical to other 486 processors but optimizing it for mobile use.

AM486DX Series (1994–1995)
All the above chips were made by Intel but AMD made compatible chips. This means their internal architecture was not the same but could run the same machine code. They used the same numbering, i.e. 486, that led Intel to use the name Pentium for their next series of chips.

AMD AM5x86 (1995)
The 5x86 ran at an effective speed of 133 MHz but could work with 33 MHz motherboards and the then new 33 MHz PCI bus. The result was a chip that was faster than Intel's Pentium 75.

The Pentium (1993)
Initially running at 60 MHz, the Pentium could achieve 100 MIPS. It was also known as the 'P5'. It had 3.21 million transistors and a 32-bit address bus like the 486 but a 64-bit external data bus, about twice the speed of the 486.

The Pentium was eventually to become available in 60, 66, 75, 90, 100, 120, 133, 150, 166 and 200 MHz versions. The first ones fitted Socket 4 boards while the rest fitted Socket 7 boards. The Pentium was 'superscalar', it could execute two instructions per clock cycle.

With two separate 8K caches it was much faster than a 486 with the same clock speed.

The Pentium Pro (1995–1999)

Detailed changes over the Pentium were made to make the Pentium Pro run faster for the same clock speeds. Three instead of two instructions can be decoded in each clock cycle. Instruction decoding and execution are decoupled, meaning that instructions can still be executed if one pipeline stops. Instructions could be executed out of order.

It has an 8K L1 cache for data and another one for instructions, and up to 1 MB of onboard L2 cache which increased performance. Also known as the 'PPro' it was optimized for 32-bit code, so it will run 16-bit code no faster than a Pentium.

Cyrix 6x86 Series (1995)

Cyrix is another chip maker that competes with Intel. Released in 1995, their 6x86 was designed as a direct competitor for the Pentium. It was known as the 'M1' and contained two super-pipelined integer units, an on-chip FPU and 16 KB of cache. Cyrix used a P-rating system, e.g. PR-120, 133, 150, 166 and 200 versions that implied better performance than the corresponding Pentium chip.

MediaGX (1996)

Made by Cyrix, this MediaGX chip was Cyrix's chip for low cost PCs. It had integrated audio and video circuitry, and other circuitry usually found on the motherboard itself. It did not fit 'standard' Socket 7 boards so did not catch on.

AMD K5 (1996)

AMD released the K5 in 1996 to compete with the Pentium. It fitted Socket 7 motherboards and was compatible with all x86 software. This chip also used the P-rating system to allow comparison with Pentium chips. It contained 24 KB of L1 cache and 4.3 million transistors.

Pentium MMX (1997)

The Pentium MMX, released in 1997, was intended to improve multi-media performance although software had to be specially written for it to have an effect. This software had to make use of the new MMX instruction set that was an extension of the normal 8086 instruction set. Other improvements produced a chip that could run faster than previous Pentiums.

AMD K6 (1997)
Available in 166 MHz to 300 MHz versions, the K6 gave performance comparable with Pentium II chips. It fitted Socket 7 boards so was a direct Pentium alternative, it could also run the MMX instruction set.

Cyrix 6x86MX (1997)
Cyrix's 6x86MX, also called the 'M2', could run the MMX instruction set. The fastest chips ran 333 MHz, or PR-466, implying a speed equivalent to a Pentium 466 MHz.

Pentium II (1997)
The Pentium II is optimized for 32-bit applications and will run the MMX instruction set. The Pentium II has 32 KB of L1 cache (16 KB each for data and instructions) and has 512 KB of L2 cache on package. This was the first chip to make use of 'Slot 1' in place of the sockets used before to prevent competitors from making direct replacement chips. Intel patented Slot 1.

Celeron (1998)
This is a cut down version of Pentium II aimed at the laptop market. It was slower as the L2 cache had been removed. Later versions such as the 300a came with 128 KB of L2 cache on board. These chips with the L2 cache performed well and it became popular to 'overclock' them to give more speed. Overclocking means to increase the clock speed, a practice not recommended by the makers. The Celeron is available in Slot 1 and Socket 370 formats.

AMD K6-2 and K6-3 (1998)
In 1998, AMD released the K6-2. With a larger L1 cache, 256 KB on-die L2 cache and Socket 7, the K6-2 sold very well.

Pentium III (1999)
The Pentium III was released in February1999 and was available in a 450 MHz version supporting a 100 MHz bus. It supported an extension to the MMX instruction set, called the SSE, aimed at further improving multimedia performance, especially 3D applications.

These chips include an integrated 'processor serial number' (PSN) seen by many as an invasion of privacy as this number could be read remotely. Some versions of the Pentium III support a 133 MHz front-side bus.

AMD Athlon (1999)
The AMD Athlon processor was released in 1999 and offers very high speeds. It has a super-pipelined, superscalar microarchitecture, nine

execution pipelines and a super-pipelined FPU. It fits into Slot A, a design that makes it easy for motherboard makers to change from a Slot 1 design to one that suits the Athlon chip's Slot A. These chips can use a 200 MHz bus called the Alpha EV6 from Digital Equipment Corporation. In May 2001, AMD released the Athlon 'Palomino', also called the Athlon 4, later renamed the Athlon XP. These chips operate at a slower clock speed than implied by the model numbers. The Athlon XP 1600+ performs at 1.4 GHz, but the average buyers will think it runs at 1.6 GHz. It should be clearly understood that clock speeds were never a good measure of chip performance but now, more than ever, clock speeds should not be looked upon as giving any useful information apart from marketing appeal.

Celeron II (2000)
This chip is an enhanced Celeron available from 533 MHz to 1.1 GHz. When the 800 MHz version came out, it could support the newer 100 MHz bus.

Duron (2000)
This chip has a 128 KB L1 cache, and 64 KB of on-die L2. Unlike the Celeron, it also works with the EV6 bus.

Pentium 4 (2000)
The Pentium 4 processor is available at speeds ranging from 1.70 GHz to 2.80 GHz but the top speed is set to increase to beyond 3.6 GHz. Bus speeds of 533 MHz and 400 MHz are available, the 533 MHz version giving data transfers of 4.2 GB/s compared with 1.06 GB/s from the Pentium III processor's 133 MHz system bus. It uses hyper pipelined technology, expanding the CPU pipeline from ten stages (of the P6) to 20 stages and two arithmetic logic units that operate at twice the speed of the processor. The chip also has an execution trace cache. This cache holds instructions that are already decoded and ready for execution.

AMD Athlon 64 (Clawhammer)
As from November 2002, AMD changed the name of its latest chips from 'Clawhammer' to 'AMD Athlon 64'. These chips will run 64- and 32-bit code simultaneously. The AMD Athlon XP processor models 2800+ and 2700+ have a 333 MHz front-side bus speed of 2.7 GB/s.

I.1 Microprocessor sockets

Socket 1 Found on 486 motherboards and supports 486 chips, plus the DX2, DX4 Overdrive.

Socket 2	Is an upgrade of Socket 1. It has 238 pins and suits the 486 chip but can support a Pentium Overdrive.
Socket 3	Similar to Socket 2 but contains 237 pins. It operates at 5 volts but can run at a switchable 3.3 volts.
Socket 4	Operating at 5 volts, Socket 4 supports the older, slower Pentium 60-66 and the Overdrive because these chips are the only Pentiums operating at 5 volts.
Socket 5	Socket 5 operates at 3.3 volts to support Pentium chips from 75 MHz to 133 MHz. Newer chips will not fit because they need an extra pin. Socket 5 has been replaced by Socket 7 although there are socket converters that allow Socket 7 processors in these Socket 5 boards.
Socket 6	Socket 6 is only a slightly more advanced Socket 3 with 235 pins and 3.3 volt operation to suit some 486 chips.
Socket 7	Operating at 2.5–3.3 volts, Socket 7 is perhaps the most common motherboard socket still in use. Although modern machines use slots for the latest microprocessors, there are plenty of Socket 7 boards still giving useful service. It supports Pentium chips from 75 MHz and above, MMX processors, the AMD K5, K6, K6-2, K6-3, 6x86, M2 and M3, and Pentium MMX Overdrives. This socket was the industry standard being suitable for sixth-generation chips by IDT, AMD and Cyrix. Intel abandoned the socket for its sixth-generation lineup in favour of Slot 1.
Socket 8	Socket 8 is used for the Pentium Pro unlike other modern Pentiums that use slots. Not common.
Slot 1	Slot 1 is used mainly for the P2, P3 and Celeron, but Pentium Pro can be fitted by using a Socket 8 on a daughtercard which is then fitted into Slot 1.
Slot 2	Slot 2 is a 330-pin version of Slot 1. The Slot 2 design allows the CPU to communicate with the L2 cache at the CPU's full clock speed, in contrast to Slot 1 which communicates at half that speed.
Slot A	Similar to Slot 1, this design suits the AMD Athlon processor. It uses a different bus protocol, called EV6, giving a 200 MHz front-side bus (FSB).
Socket 370	Socket 370 is a Socket 7 with an extra row of pins on all four sides. It is used for Pentium III. Celeron and Celeron II chips.
Socket 462	Socket 462 is also known as Socket A and is used for AMD's Athlon and Duron processors. It supports the 200 MHz EV6 bus, as well as the new 266 MHz EV6 bus.
Socket 423/478	Socket 423 is the older socket of the Pentium 4 sockets, Socket 478 supports the newer 478-pin Pentium 4s.

I.2 Summary of processor socket types

Maker	Name	Core	Socket	Process μm	Transistors (millions)	Voltage	Speeds MHz	Effective FSB speed MHz	L2 cache	Internal bus	Introduced
Intel	Pentium	P5	Socket 4	0.8	3.1	5	60–66	60–66		64 bit	Mar 1993
Intel	Pentium	P54C	Socket 5	0.6	3.2	3.38–3.52	75–120	60–66		64 bit	Mar 1994
Intel	Pentium	P54C	Socket 7	0.35	3.3	3.38–3.52	120–200	60–66		64 bit	Mar 1995
Cyrix/IBM	6x86	M1(R)	Socket 7	0.65–0.44	3	3.3–3.52	PR90-PR200	40–75		64 bit	Oct 1995
AMD	K5	Model 0–3	Socket 7	0.35	4.3	3.52	PR75-PR166	60–66		64 bit	Jun 1996
Cyrix/IBM	6x86L	M1L	Socket 7	0.35	3	2.8	PR120-PR200	50–75		64 bit	Jan 1997
Intel	Pentium MMX	P55C	Socket 7	0.35	4.5	2.8	133–233	60–66		64 bit	Jan 1997
AMD	K6	Model 6	Socket 7	0.35	8.8	2.9–3.3	166–233	66		64 bit	Apr 1997
Cyrix/IBM	6x86MX/MII	M2	Socket 7	0.35–0.25	6.6	2.9	PR166-PR366	66–83		64 bit	May 1997
AMD	K6	Model 7	Socket 7	0.25	8.8	2.2	200–300	66		64 bit	Jan 1998
AMD	K6–2 3D	Model 8/[7:0]	Socket 7	0.25	9.3	2.2	266–400	66–100		64 bit	May 1998
AMD	K6–2 3D CXT	Model 8/[F:8]	Socket 7	0.25	9.3	2.2–2.4	333–550	95–100		64 bit	Nov 1998
AMD	K6-III	Model 9	Socket 7	0.25	21.3	2.4	400–450	100	256 KB 4-way	64 bit	Feb 1999
AMD	K6–2+		Socket 7	0.18	15	2	450–550	100	128 KB	64 bit	Apr 2000
AMD	K6-III+	Model 13	Socket 7	0.18	21.3	2	450–500	95–100	256 KB 4-way	64 bit	Apr 2000
Intel	Pentium Pro	P6	Socket 8	0.6–0.35	5.5	3.1–3.3	150–200	60/66	256 KB, 512 KB, 1024 KB	64 bit	Nov 1995
Intel	Pentium II	Klamath	Slot 1	0.35	7.5	2.8	233–300	66	512 KB	64 bit	May 1997
Intel	Celeron	Covington	Slot 1	0.25	7.5	2	266–300	66		64 bit	Apr 1998
Intel	Pentium II Xeon	Drake	Slot 2	0.25	7.5	2	400–450	100	512 KB, 1024 KB, 2048 KB	64 bit	Jun 1998

continued

Maker	Name	Core	Socket	Process μm	Transistors (millions)	Voltage	Speeds MHz	Effective FSB speed MHz	L2 cache	Internal bus	Introduced
Intel	Pentium II	Deschutes	Slot 1	0.25	7.5	2	333–450	66100	512 KB	64 bit	Sep 1998
Intel	Celeron	Mendocino	Slot 1/ Socket 370	0.25	7.5 (+11.5)	2	300–533	66	128 KB	64 bit	Aug 1998
Intel	Pentium III	Katmai	Slot 1	0.18–0.25	9.5	1.65–2.05	450–600	100/133	512 KB	64 bit	Feb 1999
Intel	Pentium III Xeon	Tanner	Slot 2	0.25	9.5	2	500–550	100	512 KB, 1024 KB, 2048 KB	64 bit	Mar 1999
Intel	Pentium III	Coppermine	Slot 1/ Socket 370	0.18	12 (+16)	1.6–1.8	500–1133	100/133	256 KB	256 bit	Oct 1999
Intel	Pentium III Xeon	Cascades	Slot 2	0.18	12 (+16–128)	2.8	600–1000	100/133	256 KB, 1024 KB, 2048 KB	256 bit	Oct 1999
Intel	Celeron II	Coppermine	Socket 370	0.18	12 (+8)	1.5–1.7	533–1100	66/100	128 KB	256 bit	Mar 2000
Intel	Pentium III Server	Tualatin	Socket 370	0.13		1.1/1.45	700–1400	100/133	512 KB	256 bit	Jun 2001
Intel	Pentium III Desktop	Tualatin	Socket 370	0.13		1.45	1133–1200	133	256 KB	256 bit	Aug 2001
Intel	Pentium III Celeron	Tualatin	Socket 370	0.13		1.45	1000–1400	100	256 KB	256 bit	Oct 2001
AMD	Athlon	K7	Slot A	0.25	22	1.6	500–700	200	512 KB	64 bit	Aug 1999
AMD	Athlon	K75	Slot A	0.18	22	1.6–1.8	500–1GHz	200	512 KB	64 bit	Jan 2000
AMD	Duron	Spitfire	Socket A	0.18	25	1.5–1.6	600–950	200	64 KB	64 bit	Jun 2000
AMD	Athlon	Thunderbird	Slot A/ Socket A	0.18	22 (+15)	1.75	650–1400	200/266	256 KB	64 bit	Jun 2000
AMD	Duron	Morgan	Socket A	0.18	25	1.75	1000–1300	200	64 KB	64 bit	Aug 2001

AMD	Athlon XP	Palomina	Socket A	0.18	37.5	1.75	1333–1733 (XP1500+-XP2100+)	266	256 KB	64 bit	Oct 2001
AMD	Athlon XP	Thoroughbred A	Socket A	0.13	37.2	1.5–1.65	1467–1800 (XP1700+-XP2200+)	266	256 KB	64 bit	Jun 2002
AMD	Athlon XP	Thoroughbred B	Socket A	0.13	37.6	1.65	1800–2250 (XP2200+-XP2800+)	266–322	256 KB	64 bit	Aug 2002
Intel	Pentium 4	Willamette	Socket 423/ Socket 478	0.18	42	1.75	1300–2000	400	256 KB	256 bit	Nov 2000
Intel	Pentium 4 Celeron	Willamette	Socket 478	0.18	42	1.75	1700–1800	400	128 KB	256 bit	May 2002
Intel	Pentium 4	Northwood	Socket 478	0.13	55	1.5	1600–2533	400–533	512 KB	256 bit	Jan 2002
Intel	Pentium 4	Northwood	Socket 478	0.13	55	1.5–1.525	2500–2800	400–533	512 KB	256 bit	Aug 2002
Intel	Pentium 4 Celeron	Northwood	Socket 478	0.13	55	1.5	2000	400	128 KB	256 bit	Oct 2002
Intel	Pentium 4	Northwood	Socket 478	0.13	??	1.5	1.4–2.6 GHz	400	512 KB	256 bit	Aug 2002
Intel	Pentium 4	Northwood 'A'	Socket 478	0.13	??	1.5	2.26–3.60 GHz	533	512 KB	256 bit	Aug 2002
Intel	Pentium 4	Prescott	Socket 478	0.09	??	1.5	3.60 GHz–5.?? GHz	664	512 KB	256 bit	Mid 2003?
AMD	Athlon??	Hammer	Socket 754	0.13	??	??	??	??	??	??	Mid 2003?

Index